IDEAS
OF
STATISTICS

IDEAS OF STATISTICS

J. Leroy Folks
OKLAHOMA STATE UNIVERSITY

John Wiley & Sons
NEW YORK CHICHESTER
BRISBANE TORONTO

Photo Research by Marge Graham.
Photo Editor, Stella Kupferberg.

Library of Congress Cataloging in Publication Data

Folks, Leroy, 1929–
Ideas of statistics.

Includes index.
1. Statistics. I. Title.
QA276.12.F64 519.5 80-14723
ISBN 0-471-02099-0

Printed in the United States of America

10 9 8 7 6 5 4 3 2 1

TO MY FAMILY

PREFACE

I was first motivated to write this book several years ago when I proposed to a committee at our university that our sophomore statistics course be placed on the list of recommended courses for the general studies program. I was dismayed to discover that some committee members found it inconceivable that such a course could be necessary in this program, which makes up the liberal education part of a bachelor's degree at our university. Upon reflection I decided that the committee may have been correct; the way in which we were teaching the course was not suitable for a liberal education degree.

My second motivation occurred in the fall of 1974, when I was teaching sophomore statistics. We were using a well-known elementary statistics book that accomplishes almost all that the author intends; however, it is a book in mathematical statistics with almost all of the mathematics stripped out of it. As a result, it is difficult for students to perceive the value of the study of statistics.

This book is not subservient to mathematics; although the level of mathematics is elementary, the book seriously discusses statistical ideas. Students are generally intelligent individuals, even though some of them have little mathematical background. This book, I hope, provides an intellectual experience at the level of the student. It is suitable for a liberal education—for relating statistics to the history of people, to the humanities, to science in the broad sense, and to the world of great ideas. It is not highly structured in terms of the usual sequence of ideas, from probability to estimation and testing.

It discusses briefly the history of statistics. This includes the description of states given by Aristotle, the *Staatenkunde* of the German and Italian scholars, the political arithmetic of seventeenth-century England, the development of probability as it relates to gambling theory, and the beginnings of experimental science as it relates to the statistics of experimental design.

Information about the people involved in the development of statistics is also presented so that students will have some knowledge of Galton, Pearson, Fisher, and "Student."

Important ideas of statistics are introduced such as randomization, experimental error, the experimental principles of the Rothamsted school, the central limit theorem, the concept of an experimental design, the idea of a mathematical model on which to base statistical analyses, the idea of a conceptual population of repetitions, and the role of statistical methods in scientific research.

Students should know about the great controversies that have torn but, at the same time, enriched and solidified the profession of statistics and about the controversies surrounding Bayesian statistics that have gone on for several hundred years. A knowledge of the disagreement between Fisher and Pearson is helpful in trying to comprehend the distinction between Fisherian and Neyman-Pearson statistics.

Also, students should realize that the subject of statistics is still in the making, that the final chapter has not yet been written, that statisticians above all others do not fully understand how to collect or organize data.

The organization of the book may strike some as unorthodox; there are thirty-six short lectures with no subtitles. However, this reflects the way in which I usually organize a one-semester course: one lecture on the normal distribution, one lecture on the Poisson, for instance. Instructors using this book can introduce their own subtitles.

There is adequate material for a more traditional course in statistics. Such a course might proceed as follows: descriptive statistics (Lectures 4 to 6), probability (Lectures 7 and 8), probability distributions (Lectures 10 to 12), sampling distributions (Lectures 13 to 15), testing (Lecture 17), estimation (Lecture 18), correlation and regression (Lectures 20 and 23), *t* tests (Lecture 29), analysis of variance (Lecture 30), and chi-square (Lecture 34).

ACKNOWLEDGMENTS

I began and completed this manuscript at Oklahoma State University, but substantial portions of it were written while I was a visitor at University College London and Imperial College, London, from 1977 to 1978.

I am grateful to the Biometrika Trustees for permission to reprint adapted forms of various tables (Tables A.3, A.5, A.7, and A.8) from *Biometrika Tables for Statisticians*, 1958.

I am also obliged to the Literary Executor of the late Sir Ronald A. Fisher, F.R.S., to Dr. Frank Yates, F.R.S., and to Longman Group Ltd., London, for permission to reprint Tables III and VII from *Statistical Tables for Biological, Agricultural and Medical Research*, sixth edition, 1974.

I thank all those who encouraged me, and I acknowledge the many helpful comments from those who reviewed the manuscript, particularly Margaret Land, Bill Stewart, and Nitis Mukhopadhyay. I appreciate Gene E. Grogan's (Cities Service Corporation) providing typing assistance with the manuscript. Finally, I offer thanks to those who typed various versions of the manuscript, especially Cathy Scott, Margaret Bartlett, Sue Folks, and Kate Mellina.

J.L.F.

CONTENTS

LARGE AND SMALL SAMPLES

OPINIONS, CONCLUSIONS, AND DECISIONS

TWO DIMENSIONS

LEGENDRE'S PRINCIPLE OF LEAST SQUARES

STATISTICS OF EXPERIMENTAL SCIENCE

SOME RANDOMIZED EXPERIMENTS

IMPORTANT CONFLICTS OF IDEAS

MANY DIMENSIONS

ORIGINS OF STATISTICS

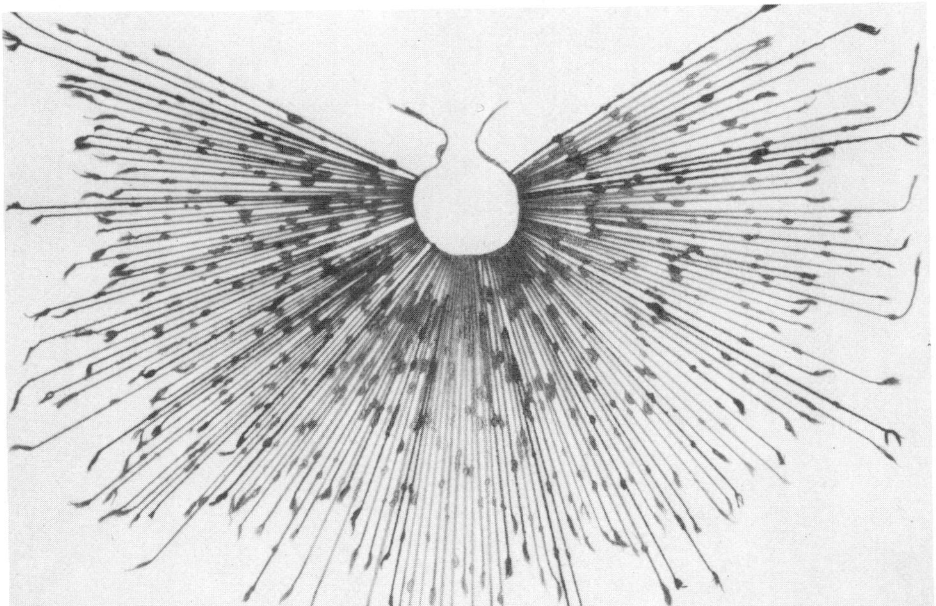

QUIPU. The ancient Inca system of recording vital statistics. Courtesy of the American Museum of Natural History.

1

POLITICAL ARITHMETIC

1.1 *And the Lord spake unto Moses in the wilderness of Sinai, in the tabernacle of the congregation, on the first day of the second month, in the second year after they were come out of the land of Egypt, saying, Take ye the sum of all the congregation to the children of Israel, after their families, by the house of their fathers, with the number of their names, every male by their polls; From twenty years old and upward, all that are able to go forth to war in Israel; thou and Aaron shall number them by their armies.*

The fourth book of the Old Testament begins with this instruction to Moses to conduct a **census** of the fighting men of Israel. In the passages that follow we are given the results of that early census, conducted about 1500 B.C. Actually, we know that censuses were carried out in much earlier times for purposes of taxation. Censuses in ancient Babylonia, China, and Egypt apparently were taken as early as 3000 B.C.

1.2 One of the most interesting accounts of an early census is given in the twenty-fourth chapter of the Book of Second Samuel. "And again the anger of the Lord was kindled against Israel and He moved David against them to say, Go, number Israel and Judah." David instructed a reluctant Joab to make a census of the people to determine the number of fighting men. It is recorded that because David did this, divine wrath was visited on Israel and that 70,000 men died of a pestilence.

The census of King David (c. 1500 B.C.) and the resulting punishment seem to have provided a basis for future public resistance of censuses. Governor Hunter of New York reported in 1712 [2], "I have issued out orders to the several counties and cities for an account, of the numbers of their inhabitants and slaves but have never been able to obtain it compleat, the people being deterr'd by a simple superstition and observation, that the sickness follow'd upon the last numbering of the people."* When we consider that the early censuses were precursors of military drafts and tax collectors, it is not surprising that they were resisted by the populace.

1.3 The word census itself is derived from the Latin word **censere**, which means to tax. The Roman census was established by the sixth king of Rome, Servius Tullius (534–378 B.C.). Under this system Roman officials called **censors** made a register at 5-year intervals of the people and their property for taxation purposes and for determining the number of able-bodied fighting men [4]. In 5 B.C. Caesar Augustus extended the census to include the entire Roman Empire. Thus it is that we have the beginning verse in the beautiful and traditional Christmas story: "And it came to pass in those days, that there went out a decree from Caesar Augustus, that all the world should be taxed." To register for such a taxation, Joseph and Mary journeyed to Bethlehem, where the infant Jesus was born. The last regular Roman census was conducted in 74 A.D. With the collapse of the Roman Empire, regular periodic censuses were not conducted in the Western world until the seventeenth century.

1.4 The name **statistics** can be traced to the Latin words **status**, meaning state, and **statista**, meaning statesman. Aristotle (384–322 B.C.) [5] was born in Macedonia, studied under Plato in Athens, and was a tutor to Alexander at the request of King Phillip. He established his own school in Athens when Alexander inherited the throne. The *Politeiai* of Aristotle contained a description of 158 states. This initial attempt at the **comparative** description of states was subsequently developed by Italian and German authors into a subject called statistics (**Staatenkunde** in German). Westergaard [6] traces the development of the description of states.

1.5 During the Middle Ages the system of feudalism more or less rendered national censuses impossible, although there were attempts to revive them. One notable example is the breviary of Charlemagne in 808 A.D..

At Christmas in 1085, William the Conqueror ordered a statistical survey of England. The record of this survey is contained in the

* From E. B. O'Callaghan, ed. *Documents Relative to the Colonial History of the State of New York*, Vol. V (Albany, N.Y.: Weed, Parsons, & Co., 1855, p. 339).

Domesday Book [4]. The survey collected information of land, landowners, land use, tenants and servants, and livestock and served as the basis for taxes until 1522 when a new *Domesday Book* was completed.

1.6 Early in the sixteenth century **Bills of Mortality** (published summaries) began to appear in London. David [3] traces the beginning of these bills to an order from Thomas Cromwell, acting on behalf of Henry VIII. It has been speculated that the king desired these summaries because of his great fear of the plague. However, Cassedy [2] indicates that it is very difficult to determine the beginning of the bills and that the precise date is unknown. In the beginning the Bills of Mortality recorded only deaths from the plague. Over the years they were expanded to include christenings and, around the end of the sixteenth century, data on deaths from other diseases.

1.7 The Spaniards conducted very early censuses in the Americas. A census of Peru in 1548 was carried out by the Spanish viceroy, Don Pedro de la Fasca. This census is described by Carlos A. Uriarte in the March 1949 issue of *Estadistica.* Before the Spaniards came, the Incas had their own system of recording statistics. This system used intertwined colored strings and knots known as **quipus**. Cassedy [2, p. 3] quotes historian William H. Prescott in his description of the system of **quipus** as a method "which has scarcely a counterpart in the annals of a semicivilized people. A register was kept of all the births and deaths throughout the country and exact returns of the actual population were made to government every year by means of the **quipus**."*

1.8 In seventeenth-century England there was great interest in the so-called **political arithmetic**, which consisted largely of analyses of recorded births and deaths. In 1662 John Graunt published his first and only book, a remarkable manuscript entitled *Natural and Political Observations upon the Bills of Mortality.* Despite the unreliable nature of the data contained in the Bills of Mortality, Graunt made an exhaustive study of the information contained therein and noted many regularities and irregularities. For example, he noted that the fraction of male births is almost exactly that of female births, the fraction of male births being slightly greater. This observation, well known in our day, seems to have been new and surprising in 1662. Over a long series of years, he counted

* Quoted in *Handbook of Vital Statistics* (N.Y. Statistical Office of the United Nations, 1955, p. 4).

the christenings of 139,782 boys compared with 130,866 girls. Graunt made a determined if awkward attempt to develop a mortality table of the type used by insurance companies today.

1.9 The word **statistics** was coined by the German scholar Gottfried Achenwall about the middle of the eighteenth century. Of course, it was derived from the word status and the German counterpart of political arithmetic. The word was apparently used for the first time in Great Britain by Sir John Sinclair who, in a series of volumes published between 1791 and 1799, gave a statistical account of Scotland drawn up from the communications of the ministers of the different parishes. Yule [7] quotes Sinclair as saying, "Many people were at first surprised at my using the new words Statistics and Statistical, as it was supposed that some term in our own language might have expressed the same meaning."* It is hard to believe today that such a short time ago the word statistics was considered new.

Boorstin [1] points out that **statistics** appeared in the *Encyclopaedia Britannica* in 1797. He also mentions that for a time the word **publicistics** competed in literary use. It is interesting to speculate about the course of events if publicistics had won. Would there have been an American Publicistical Association or a Royal Publicistical Society?

1.10 When the Constitution of the United States was written, the census became a regular and vital part of the government. The census was provided for in Article I, Section 2, which states, "The actual enumeration shall be made within three years after the first meeting of the Congress of the United States, and within every subsequent term of ten years, in such manner as they shall by law direct." The first decennial census in the United States was taken in 1790, and other censuses have followed every 10 years since. In addition, censuses in certain fields have been taken at more frequent intervals.

1.11 In this lecture we have sketched a few of the origins of census data and the analysis of such data. The subject of statistics, which owes its name to the description of states, has expanded far beyond these original boundaries, but the modern versions of political arithmetic and description of states constitute an important part of statistics.

* Reproduced by permission of the publishers, Charles Griffin & Company, Ltd., of London and High Wycombe, from Yule, *An Introduction to the Theory of Statistics*, 5th Ed., 1919 (15th Ed., Yule & Kendall, 1950).

SUMMARY. The root word of statistics is **state**, and the description of states constitutes one of the important roots of modern statistics. Partial descriptions of nations and states were provided by the early **censuses**, which were conducted in order to raise armies and establish tax rolls. Many examples of censuses can be found in the Old Testament, and the New Testament begins with the account of a census of the Roman Empire conducted by Caesar Augustus.

German and Italian scholars developed the description of states into a subject that more nearly resembles modern statistics. Under William the Conqueror, a detailed statistical survey was made of England and recorded in the *Domesday Book*. In seventeenth-century England **political arithmetic** flourished. Consisting largely of analyses of recorded births and deaths, this provided some of the first **mortality** tables.

The census was required by the Constitution of the United States, although the word statistics itself was not firmly established until 1797, when it first appeared in the *Encyclopaedia Britannica*.

REFERENCES

1. Boorstin, Daniel. (1973). *The Americans: The Democratic Experience*, New York: Random House.
2. Cassedy, James H. (1969). *Demography in Early America*, Cambridge, Mass.: Harvard University Press.
3. David, F. N. (1962). *Games, Gods and Gambling*, New York: Hafner.
4. Dudley, Lavinia P., Exec. Ed. (1957). *Encyclopedia Americana*, New York: Americana Corporation.
5. Farrington, Benjamin. (1965). *Aristotle, Founder of Scientific Philosophy*, London: Weidenfeld & Nicolson.
6. Westergaard, Harald. (1968). *Contributions to the History of Statistics*, New York: Agathon Press.
7. Yule, G. Udny. (1919). *An Introduction to the Theory of Statistics*, London: Charles Griffin.

EXERCISES

1. Give a summary of the statistical information available from the U.S. Census Bureau.

2. Give a summary of statistical information based on sampling provided by the U.S. Census Bureau.
3. What was the population of the United States as determined by the 1790 census?
4. Find other censuses referred to in the Bible other than those mentioned in the lecture.
5. Try to determine if your hometown has any local statistical record.
6. Try to find characteristics of different countries that are not described statistically.
7. The *Statistical Abstract* of the United States gives detailed information on many characteristics of the United States. Using the most recent edition, what would you say is the windiest city? The hottest city? The sunniest city? What qualifications do you wish to attach to your answer?
8. Using the 1977 *County and City Data Book* issued by the U.S. Bureau of the Census, take a random sample of 10 counties from your state. Estimate the percent unemployed for the state using only the county figures. How close is your estimate to the figure reported for the state?
9. Using data from the *Statistical Abstract*, graph crude oil production in the United States from 1950 to the present.
10. To attack the major problems confronting humanity (hunger, war, depletion of raw materials, disease, pollution, etc.), can you think of any statistical information that should be maintained and is not?

BLAISE PASCAL. 1623–1662. Courtesy of IBM.

2

THE CASINO

2.1 The **random event** came slowly to Western civilization. **Games of chance** were played thousands of years before Christ and continued to be played in times dated A.D., despite vigorous opposition from the Church. Yet explicit formulation of the concept of a random event and the subsequent theory of **probability** came late; some historians date the beginnings of probability theory as recently as 1500 A.D.

2.2 The **astragalus** is a bone that lies above the **talus**, or heelbone. In ancient and medieval literature, the words **astragalus**, **talus**, **knucklebone**, and **hucklebone** were used indiscriminately. Interesting and scholarly accounts of the use of the astragalus in early games of chance are given by David [2, 3]. We know from archaeological findings that large numbers of astragali, particularly from hooved animals, were collected in ancient times. We also know that these bones were used in games of various kinds. As pointed out by David, the astragalus was certainly used in various board games in Egypt (c. 3500 B.C.); a game called "hounds and jackals" by the excavators of Egyptian tombs apparently used astragali in the same manner as dice. The hounds and jackals were moved according to the results of throwing the astragali. David notes that Homer (c. 900 B.C.) tells us that Patroclus became angry with his opponent while playing a game of knucklebones and nearly killed him.

By the time of the Romans, gaming was very popular and laws were passed forbidding and regulating it. Stern opposition by the Church followed and continues today but, despite all such opposition, gaming and gambling have flourished among all classes of people. In the early days most games involved the astragalus; **dice** and **cards** were used later.

According to David, the **earliest** dice found so far date from the beginning of the third millenium. A die found in northern Iraq and made of well-fired buff pottery has the opposite points in consecutive order: 2 opposite 3, 4 opposite 5, and 6 opposite 1. David suggests that the die with opposite faces totaling 7 must have evolved about 1400 B.C.

In addition to the use of games, **chance mechanisms** have been used throughout the ages to divine the will of the gods. David records many such practices and gives an interesting account of such a practice in fairly recent times. As recently as 1737, John Wesley sought guidance by resorting to the practice of **drawing lots** to decide whether or not to marry.

2.3 Although references to games of chance are numerous and span thousands of years, elementary concepts of probability appear in the literature only in the last few hundred years. Why was the theory of probability so slow in being developed? In an article on the history of probability, Kendall [4] raises this question and offers an opinion. He suggests that it was because of moral and religious barriers. It certainly seems plausible that the opposition of the Church may have been based on theological implications and not on social consequences of gambling. However, the explanation offered by Kendall remains speculative.

H. Walker [8] refers to a reference given by Smith [6] that suggests some idea of probability in Chinese literature as early as 220 B.C.

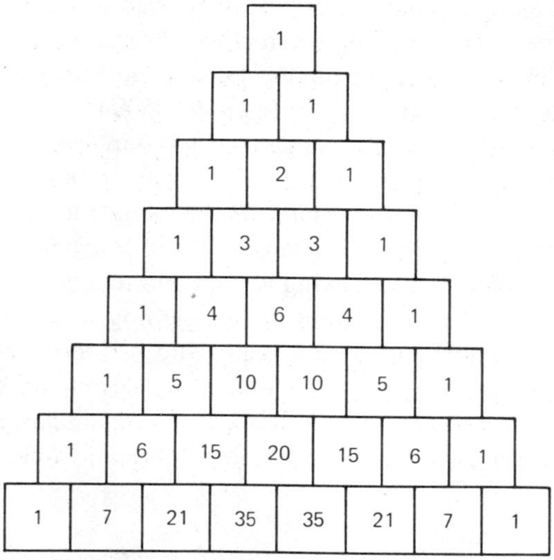

Figure 2.1 Pascal's Arithmetical Triangle.

Kendall draws our attention to an early example of counting the possibilities in dice games. Bishop Wibold, about 960 A.D., enumerated 56 virtues, letting each virtue correspond to the outcome of the throw of three dice. When three dice are thrown, there are 56 possible outcomes when we ignore the order (e.g., 2, 2, 1 is the same as 1, 2, 2). The dice were then thrown, and the thrower concentrated on the corresponding virtue (honesty, loyalty, etc.) for some time. There are many references to counting problems associated with games of chance prior to the development of ideas of probability. David references the Chinese writer Chu Shih-chieh, who published the **arithmetical triangle** of binomial coefficients in 1303 while describing it as an ancient method. He gives a diagram similar to that in Figure 2.1. This triangle array is known today as **Pascal's triangle**, although it was published by Pascal 351 years after Shih-chieh. We defer a fuller explanation until later, when we will have several occasions to make use of this triangle.

2.4 One of the earliest examples of a gambling problem in a mathematical work is given in the work of Fra Luca Pacciolo called *Summa de arithmetica, geometrica, proportioni e proportionalita*, which was printed in Venice in 1494. He gives the problem of how to divide equitably the stakes between two players when a game is interrupted before its conclusion. This is one of the earliest versions of **the problem of points**. Walker mentions that this problem was repeated in almost all subsequent works on probability.

Although not published until 1663, the first work containing extensive material on probability is *The Book on Games of Chance* (*Liber de lido alae*) by Cardano [1]. Cardano wrote this book about 1520 while he was at the University of Padua. In addition to being a handbook on gambling, it represents one of the first examples of putting the odds for chance events into mathematical terms. Although he made numerous errors, Cardano solved many problems, formulating them in terms very similar to those of today.

2.5 It is of interest that Kepler made some remarks on chance in a work published in 1606 (see Todhunter [7, p. 4]). Kepler was speculating about the cause of the appearance of a new star that shone brightly in 1604. Todhunter also mentions that Galileo made some contributions to probability in a work published in 1642. Galileo solved a problem involving the throw of three dice. He showed that of the 216 possible cases, there are 27 that give the number 10 and 25 that give the number 9.

2.6 The next major recorded work in probability is found in correspondence between Blaise Pascal and Pierre de Fermat. Pascal and Fermat were two of the most distinguished mathematicians of their time. Pascal (1623–1662) established lasting fame as a physicist and mathematician by the age of 25, when he retired to a life of religious contemplation. Fermat (1608–1665) stated a number of remarkable propositions in the theory of numbers. Todhunter notes, "Neglecting the trifling hints which may be found in preceding writers we may say that the Theory of Probability really commenced with Pascal and Fermat; and it would be difficult to find two names which could confer higher honour on the subject."*

The Chevalier de Mere is famous for having posed questions that started the Pascal-Fermat correspondence, although little is known about de Mere. David says that "It has never been stated how Pascal and de Mere began to discuss such problems but it may be noted that this contact was made during what might be called Pascal's dissolute period. Since de Mere was not a member of the continuance of what has been the Mersenne Academy, they may have met in less academic surroundings."† The speculation is frequently made that the Chevalier's questions were motivated by his occasional visits at the casino. Apparently de Mere posed the **problem of points** to Pascal. Suppose a gambler is undertaking to throw a six with a die in eight throws. If he does not succeed in the first three throws, what proportion of the stake should he be entitled to if he gives up his fourth throw? Pascal and Fermat carried on an extensive correspondence about this and other problems of points.

Among the various dice games of that age was the following. The house offers to bet even money that in four throws of a single die at least one six will occur. It is claimed that de Mere was interested in a slightly more complicated game. Suppose that two dice were thrown. The thinking of the day argued that since two dice can result in six times as many outcomes as one die, 24 (6×4) throws of two dice should be favorable for the house that at least one pair of double sixes will occur. It is believed by some that de Mere had obtained empirical evidence (in the casino, no doubt) that in 24 throws the odds were against the house. Pascal showed that the probability of at least one double six was, in fact, equal to .491.

* Reprinted with permission from I. Todhunter, *A History of the Mathematical Theory of Probability* (Chelsea Publishing Company, 1949).

† Reproduced by permission of the publishers, Charles Griffin & Company Ltd, of London and High Wycombe, from David, *Games, Gods, and Gambling*, 1962.

Another item in the Pascal-Fermat correspondence is the arithmetical triangle referred to earlier. Although Pascal was one of the last of a long line of discoverers of this device, it is interesting that it is generally known as Pascal's triangle.

2.7 From the seventeenth century to the present, the theory of probability has developed far beyond its simple origins of dealing with games of chance. The story of these years is unbelievably rich, and we will conclude this lecture with only a few fragments. The story includes the work of Leibnitz, published in 1666, in which a table similar to the arithmetical triangle is given. Leibnitz gives a curious notation. When a set of things is taken two at a time, he uses com2natio; when three at a time com3natio; when four at a time com4natio; etc. Leibnitz used $+$, $-$, $=$ in their present sense. He denotes multiplication and division by $\cap$ and $\cup$, which are used in modern set theory versions of multiplication and addition. Curiously, he used the word productum to denote a sum, as in 8 is the productum of $5+3$.

Another version of the arithmetical triangle is given in the *Ars Conjectandi*, 1713, published posthumously after the death of its author, James Bernoulli (1654–1705). In many ways this work is extremely important. Bernoulli maintained emphatically that there was application for probability in civil, moral, and economic affairs. However, it remained for those who came later to find applications.

In a paper published on November 12, 1733 and distributed only to friends, De Moivre gave the famous **normal curve**. Walker points out that this origin of the normal curve was discovered by Pearson [8].

In the late 1700s and early 1800s, when the United States was in its infancy, the **normal curve** was being firmly established by the great mathematicians Laplace and Gauss. Gauss made use of the normal curve in the analysis of astronomical data. It is curious that the normal curve is widely known as the **Gaussian** curve or **distribution**.

Bernoulli's belief that probability would come to be applied in many fields of human endeavor has been justified. Much remains to be done, far removed from the casino, but the gambling origins remain a vital and interesting part of the subject.

SUMMARY. **Games of chance** have been played from thousands of years before Christ to the present. All sorts of **chance** mechanisms have been used, but one of the earliest involved the throwing of polished bones called **astragali**. Dice and cards came much later.

Although games of chance and chance mechanisms were well known, no adequate theory of **probability** existed. No real development of the subject even began until the middle of the **seventeenth century**. Certainly some traces of the subject can be found much earlier, but the development of probability as a serious subject began in the Western world with the work of **Pascal** and **Fermat.** Their work, directed specifically toward problems in gambling, started probability on its path.

With the work of men such as **Bernoulli** and **De Moivre** in the eighteenth century and **Gauss** and **Laplace** in the nineteenth century, the theory of probability was ready for the torrent of statistical development that began in the late nineteenth century.

REFERENCES

1. Cardano, Gerolamo. (1961). *The Book on Games of Chance*, New York: Holt, Rinehart and Winston. Translated by Sydney Henry Gould, foreword by Samuel S. Wilks, reprinted from *Cardano: The Gambling Scholar* by Oystein Ore (Princeton University Press).
2. David, F. N. (1955). "Studies in the History of Probability and Statistics," *Biometrika*, 42, 1–15.
3. David, F. N. (1962). *Games, Gods, and Gambling*, New York: Hafner.
4. Kendall, M. G. (1956). "Studies in the History of Probability and Statistics," *Biometrika*, 43, 1–14.
5. Pearson, K. (1924). "Historical Note on the Normal Curve of Errors," *Biometrika*, 16, 402–404.
6. Smith, D. E. (1925). *History of Mathematics II*, Boston: Ginn.
7. Todhunter, I. (1865). *A History of the Mathematical Theory of Probability*, New York: reprint by F. E. Stechert, 1931.
8. Walker, Helen M. (1931). *Studies in the History of Statistical Method*, Baltimore: Williams & Wilkins.

EXERCISES

1. Prepare a short report on the gambling origins of probability. Most encyclopedias contain material that will aid you.
2. Try to discover the rule for generating successive rows of Pascal's triangle.

3. Throw a pair of dice 24 times and record the number of occurrences of a double six. Repeat this exercise 20 times and record the number of times that at least one six occurs. Is this result in reasonable agreement with Pascal's result?
4. A carnival game offers a prize of $5 if you get at least one six in three tosses of a die. What would you be willing to pay for a chance to play? What do you feel are the odds in favor of your winning? After throwing the die once and not getting a six? After throwing twice without producing a six?
5. Two women, meeting again at a class reunion, agree to play the following game. The person with the least money in her wallet will lose all of her money to the other. Each is thinking, "If I lose, I will lose less than I will win, if I win. So the game is in my favor." Where is the fallacy?
6. A magazine promotion campaign sponsored a $100,000 sweepstakes. Each person receiving an entry blank was given the chance of winning the grand prize of $100,000. If the chance of winning is 1 out of 100,000, is it worth the postage stamp to enter?
7. A professional gambler wishes to purchase a pair of loaded dice. He hires an artisan to build a pair of dice for which the probability of rolling a pair of ones is exactly 1 in 64 instead of 1 in 36. He pays for the dice and is told that every possible calculation has been made and that the desired probability is undoubtedly 1/64. However, the dice have never been thrown. Would you believe the artisan's claim?

MENDEL'S GARDEN. The small monastery garden where the pea experiments were conducted. Courtesy of Professor Dr. Ing. Jaroslav Krinzenecky.

3

THE SCIENTISTS

3.1 In the spring of 1969, a series of three lectures at the Johns Hopkins University [2] dealt with an ancient theme—the relationship between **observation** and **theory**. The lecture by Ernest Nagel is entitled "Theory and Observation" and begins by discussing Einstein's views on the antithesis between the **deductive** and **empirical** components of science. Nagel quotes Einstein as saying, "... purely logical thinking cannot yield us any knowledge of the empirical world; all knowledge of reality starts from experience and ends with it."*

The logical status of scientific theories in their relation to observation has concerned scientists and philosophers for centuries. In my view, modern statistics helps to relate observation to theory and is therefore an integral and vital part of scientific thinking. However, statistical methods based on statistical theory have existed for only about 100 years. In this lecture we wish to mention a few of the people and ideas related to the development of statistics in science.

3.2 In 1620 Francis Bacon published the *Novum Organum.* In this remarkable document, Bacon makes certain pronouncements on the **scientific method**. Without direct experience in the laboratory, Bacon concludes that we can learn only through observation. "Man, as the minister and interpreter of nature, does and understands as much as his observations on the order of nature, either with regard to things or the mind permit him, and neither knows nor is capable of more"† [4, p. 37].

* From Nagel, Bromberger, and Grunbaum, *Observation and Theory in Science*, copyright © 1971 by the Johns Hopkins Press.

† From *Moments of Discovery: The Origins of Science*, ed. by George Schwartz and Philip W. Bishop, copyright © 1958 by Basic Books, Inc., Publishers, New York.

He proposed investigating all facets of the universe by collecting all possible data and performing every conceivable experiment. Then the data were to be tabulated and analyzed. Although Bacon's proposal of such an exhaustive task seems naive, it represented an important endorsement of the empirical method.

From the very beginning of science, of course, data were analyzed and interpreted. However, this did not resemble what we call statistics today until the ideas of probability and political arithmetic began to be merged in the nineteenth century.

3.3 As early as 1783, Laplace proposed the normal curve equation as being appropriate for the probability **distribution of errors.** This curve is given in Figure 3.1.

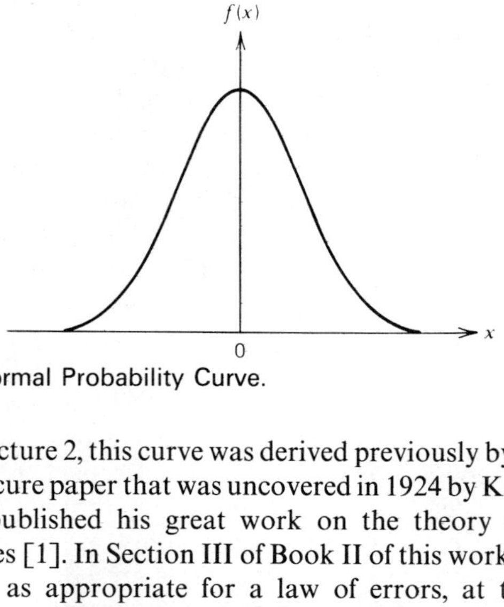

Figure 3.1 Normal Probability Curve.

As noted in Lecture 2, this curve was derived previously by De Moivre in 1733 in an obscure paper that was uncovered in 1924 by Karl Pearson. In 1809 Gauss published his great work on the theory of motions of heavenly bodies [1]. In Section III of Book II of this work he derives the normal curve as appropriate for a law of errors, at the same time acknowledging Laplace's earlier derivation. The basic idea is that if errors of observation are made up from the addition of many small, independent errors, positive errors will be equally as likely as negative errors, with the concentration close to zero, as in Figure 3.1.

With the use of the normal probability curve by Gauss and Laplace, the theory of probability was firmly launched on its way to applications in almost all fields of scientific endeavor. Bernoulli's belief that probability could have applications was beginning to be justified.

The great Belgian mathematician and astronomer Adolphe Quetelet (1796–1874) made a giant step in the development of statistics in the social sciences when he extended the use of the normal curve from errors of observation to data of various kinds. He observed that if the heights of a large number of men were shown on a bar graph, the picture resembled the normal curve. Thus he began the subject of social statistics. Although Quetelet is best known for his **moral statistics**, the effort to quantify psychological and social facts and customs, he recognized the universality of the emerging subject. Walker [6, p. 41] notes that Quetelet presented to the British Statistical Association in 1841 a list of more than 40 topics that should be investigated by statistical methods.

One of the intriguing theories of Quetelet was the theory of **l'homme moyen**, or **the average man**. Apparently he felt that the average man, if such existed, represented all that was good and beautiful.

3.4 In the latter half of the nineteenth century, statistics played a major role in the study of the laws of heredity. Darwin returned from the voyage of the *Beagle* to publish *The Origin of Species* in 1859. This work had a profound effect on many of the great minds.

In Austrian Silesia, now in Czechoslovakia, Gregor Mendel had returned to Brno as a supply teacher in July 1853. In 1856 he embarked on his experiments with the edible pea, **Pisum sativum**. In this work, which continued until 1863, Mendel discovered that the laws of heredity are statistical. He selected seven of the characteristics on which peas differ: seed shape, seed color, pod shape, color of seed coat, color of unripe pods, position of flowers, and length of stem. To explain the idea, consider color. Mendel crossed green peas with yellow peas; the result was that all the next generation were yellow. With each of the other six characteristics, he obtained similar results. The new generation of peas always resembled one of the parents.

When a second generation of peas was bred from the yellow peas of mixed parentage, about three-quarters of the peas were yellow, the others green. Mendel's data are given by Sootin [5].

Yellow Seeds	Green Seeds	Ratio of Yellow to Green Seeds
6022	2001	3.01 to 1

The agreement with a 3:1 ratio is remarkable, and a similarly strong agreement exists in Mendel's data on the other six characteristics.

Mendel presented his work at two meetings of Naturforschenden Vereins in Brno in 1865 and published the results in a paper in 1866. Curiously, his work was either unnoticed or ignored until about 1900, when it was rediscovered. The 3:1 ratio results from the following simple explanation. The pure yellow and pure green peas have two characteristics.

Pure Yellow. YY
Pure Green. gg

When these peas are crossed, there is a **random selection** of one of the characteristics from each, resulting in Yg in the first (F_1) generation. When hybrid peas from the F_1 generation are crossed again, there is a **random selection** of one of the characteristics from each of the parents, resulting in the following in the second (F_2) generation:

YY Yg gY gg

Thus Mendel was led to the 1:2:1 ratio. In the case where one is **dominant** over the other, as with yellow over green, the observed ratio is 3:1.

Mendel carried his research to the third generation, but we will not discuss that. From today's viewpoint it is hard to appreciate the magnitude of his ideas. But Mendel was not operating from such a viewpoint. His idea, which seems so simple now, must rank among the truly great ideas in history. Robert Olby, in the Author's Note to *The Origins of Mendelism* [3], quotes from a B.B.C. talk: "Often as Newton said, the successor can see further than his precursors because, standing on their shoulders, he can see a little further than they. There was nothing of this said about Gregor Mendel. He had no precursors to stand on at all."*

In England the work of Darwin had a profound effect on his cousin, Sir Francis Galton. Better equipped mathematically than Darwin to deal with a theory of evolution, Galton published many articles and books on heredity. *Hereditary Genius* was published in 1869 and *Natural Inheritance* in 1889. Galton was the first to use **correlation** and **regression**, concepts of tremendous importance that we will discuss in later lectures.

It is certain that Galton knew of the 1 : 2 : 1 Mendelian ratio and that he transmitted this ratio to Darwin in a letter written in 1875 [3, p. 72]. Galton mentions the Pascal triangle as one way of arriving at the ratio (the third line of the triangle is 1 : 2 : 1). Neither Darwin nor Galton seems to have appreciated the importance of this ratio.

3.5 At University College, London, in the 1890s Pearson, excited by Galton's *Natural Inheritance*, began to apply mathematics and the theory of probability to Darwinian evolution, and the era of modern statistics began. In 1901 the first issue of *Biometrika*, a journal devoted to the application of mathematics to biology, appeared; Pearson was one of the editors.

In the twentieth century, statistics has developed far beyond its 1901 confines and has proven to be useful in all fields of scientific endeavor.

SUMMARY. The origins of **experimental science** lie shrouded in the past. However, a few traces can be found prior to the seventeenth century. In 1620 **Francis Bacon** argued that man can learn only by **observation.** With the rapid expansion of experimental science in the eighteenth and nineteenth centuries, scientists began to make use of the existent **probability theory**.

Gauss and **Laplace** made extensive use of the normal curve in the nineteenth century, as did the great Belgian astronomer **Quetelet.**

In a monumental piece of experimental research between 1856 and 1863, **Mendel** laid bare the **statistical laws** of **genetics** and established the foundations of **Mendelian genetics**. Without any precedent, Mendel perceived that the genetic mechanism operated like a random device.

In England the work of **Darwin** influenced **Galton**, who made important contributions to the foundations of **biometry**. In turn, **Pearson**, heavily influenced by the work of Galton, launched a career that was to earn him the title of **father of statistics**.

REFERENCES

1. Gauss, Karl Friedrich. (1963). *Theory of the Motion of the Heavenly Bodies*, translation of *Theoria Notus* by Charles Henry Davis, New York: Dover.

2. Nagel, Ernest, Sylvain Bromberger, and Adolf Grunbaum. (1969). *Observation and Theory in Science*, Baltimore: The Johns Hopkins Press.
3. Olby, Robert C. (1966). *Origins of Mendelism*, London: Constable.
4. Schwartz, George, and Philip W. Bishop. (1958). *Moments of Discovery: The Origins of Science*, Vol. I, New York: Basic Books.
5. Sootin, Harry. (1959). *Gregor Mendel, Father of the Science of Genetics*, New York: Vanguard.
6. Walker, Helen M. (1931). *Studies in the History of Statistical Method*, Baltimore: Williams & Wilkins.

EXERCISES

1. Look at the textbooks and periodicals in your major field to find some use of the normal curve.
2. Find a description of the scientific method. At what stages of the scientific method do you feel statistics can enter?
3. Look at Darwin's *Origin of Species*. Did Darwin make use of any statistical arguments?
4. To what extent are theories of evolution based on statistical evidence?
5. In the science courses you have taken (high school and college) were there any scientific theories based strictly on theoretical reasoning without observations? Are you sure?
6. Simulate the pea experiment of Mendel's by tossing a pair of coins (a quarter and a nickel). Let two heads correspond to YY. Let heads on the quarter and tails on the nickel correspond to Yg; let tails on the quarter and heads on the nickel correspond to gY, and let two tails correspond to gg. Toss the pair of coins 100 times and record the number of occurrences of YY, Yg, gY, and gg. Are the results in reasonable agreement with the 1:2:1 ratio?
7. Carry on the theory of the pea breeding experiment to a third generation, crossing a yellow pea from F_2 with a green pea from F_2. What should be the ratio of yellow to green peas?

DESCRIPTION OF DATA

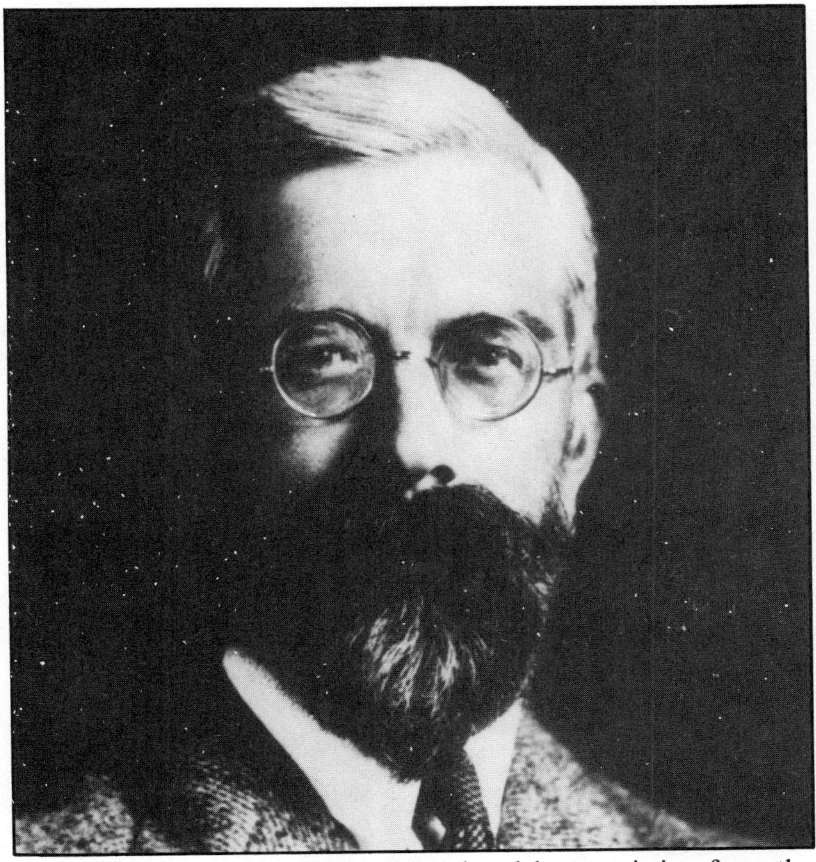

R. A. FISHER. 1890–1962. Reproduced by permission from the Department of Statistics, Iowa State University.

4

POPULATION AND SAMPLE

4.1 In 1693 Edmund Halley gave tables for "the mortality of mankind" that were obtained from mortality data for the city of Breslau. In the nineteenth century John Bennet Lawes used annual records of wheat yields on five plots at Rothamsted to estimate the change in yield per acre from 1852 to 1879 for all of England and Wales. These and other early examples of the use of **sampling** are given by Stephan [7]. They illustrate the idea of making inferences about a set of individuals, the **population**, having observed only a part of the population, the **sample**. We may have populations of plants, insects, rocks, words, books, laboratory measurements, or even populations of populations. In most cases numbers are associated with each member of the population; the statistician is usually interested in the associated population of numbers.

The distinction between population and sample represents one of the most important concepts in statistics. Making this distinction has been at the foundation of most of statistical theory and methodology since the late nineteenth century. This simple distinction was difficult to make in the early, formative years of statistics and provides no end of difficulty to today's unwary students. Much of statistics is concerned with making inferences about a population on the basis of a sample from that population. It is crucial that the distinction be made.

4.2 As the nineteenth century ended, Pearson (1857–1936) and his associates were applying statistical methods to a wide variety of problems in evolution, heredity, and related fields. Through their work, mathematical probability came to be intimately associated with statistics. A great deal of their work was involved with the mathematical description of populations, called **universes** or **aggregates** (or **kollectivs** by the German scholars). Although some of Pearson's work had to do with samples, the distinction between samples and populations was not always maintained. In the early 1900s the distinction between population and sample became better established, and Yule [9] made the distinction in his textbook in 1910.

4.3 In a very important paper that appeared in 1922 entitled "On the Mathematical Foundations of Theoretical Statistics," Fisher [2] did a great deal to solidify the concepts of population, sample, and the distinction between the two. He stated that the problems of statistics can be divided into three types.

1. **Problems of Specification.** These have to do with the choice of the **probability distribution** to describe the population.
2. **Problems of Estimation.** These involve the calculation of quantities from the sample called **statistics** to represent quantities from the population called **parameters**.
3. **Problems of Distribution.** These problems involve the **distribution of statistics** and will be discussed at length from Lecture 13 through the rest of the book.

Of course, there are other problems of statistics. For example, one of the major problems not covered by this classification is the design of experiments. This is discussed at some length in Lectures 28 to 30.

4.4 In subsequent lectures a certain amount of notation will be introduced. The distinction between quantities in the sample and corresponding quantities in the population will be maintained by adhering to the following widespread, well-established convention. Where possible, population quantities will be denoted by Greek letters and sample quantities by Roman letters. This will help students to avoid confusion.

4.5 In many subjects there is a **canonical problem** or a **standard form** that, although simplified, provides a basis for thinking and a base on which to develop more complicated ways of thinking. The canonical situation we wish to consider occurs when we have a random sample from a specified, existent population and we wish to determine from the sample the properties of the population. There are three points that we wish to make.

1. The population is an **existent** population (e.g., the population of sentences in this book).
2. The sample is a **random** sample (to be defined presently), **not** a **representative** sample (e.g., 10 sentences selected at random from the population of all sentences).
3. We may be interested in properties of the population such as the average number of words per sentence, but we have only a sample. From the sample we must make our conclusions.

We will now comment on points 2 and 3 in greater detail. There is a deeply ingrained feeling on the part of many people that the sample should be representative of the population—that it should be the population in miniature. A little thought will show that this is impossible. A sample of 8 people cannot possibly be "like" a population of 2000 people. We would need the same proportional representation in the sample as in the population for age, sex, race, education, income, etc. We might decide to choose 4 people under 30 and 4 people over 30. For each of the sets of 4 people we might then choose two males and two females. For each of the four sets of two we might choose one black and one white. To consider any of the other characteristics of the population would literally require splitting people into pieces. As appealing as the idea of a representative sample is, it has not proven to be a viable concept. There seems to be no way to give an objective definition of representative.

The idea that is now widely accepted is that of choosing the sample at random. This does not mean haphazardly, subject to the whims and prejudices of the investigator. A random sample is a sample chosen so that every **possible** sample with the same number of observations has an **equal chance** of being chosen.

The formulation of a definition of **random sample** has sometimes been credited to John Venn and C. S. Peirce (pronounced "Pers") (Carnap, [1, pp. 493–494]). Although Venn discusses randomness at length, he does not give an explicit formulation of a random sample. On the other hand, the writings of Peirce are replete with definitions of a random sample. For example, Peirce [5] says, "A sample is a **random** one, provided it is drawn by such machinery, artificial or physiological, that in the long run any one individual of the whole lot would get taken as often as any other."

The process of making conclusions from the sample about the population is called **statistical inference**. In subsequent lectures we will encounter three main areas of inference.

1. Calculating numbers called **estimates** from the sample to use as values for population constants called **parameters**.
2. Calculating **intervals** of certainty (or uncertainty) about our estimates.
3. Using **statistical tests** to judge the extent to which the sample information supports hypothesized values for the population parameters.

Peirce clearly understood the role of samples in making inferences about populations. The definition of random sample is taken from the section titled "Reasoning from Samples," and Peirce says, "The truth is that induction is reasoning from a sample taken at random to the whole lot sampled."

4.6 Once the decision has been made to choose a sample at random, there is the nontrivial problem of how to accomplish this. Conceptually, one of the simplest ways to go about it would be to number the individuals, write their numbers on tickets, and draw tickets from a hopper after stirring the tickets. Various other devices would be equivalent. However, a technique that has been in use for about 50 years is the use of **random number tables**. Instead of using a physical contraption such as a hopper with tickets or an urn with numbered balls, one simply turns to a table of random digits, reads the desired number of digits, and chooses the corresponding individuals for his sample.

Apparently the first table of random digits was published by Tippett in 1927 [8]. It was obtained by recording some of the digits taken from English tax records. Hald [4] presented 15,000 digits that were compiled from drawings in the Danish State Lottery of 1948. Fisher and Yates [3] abstracted from *Logarithmetica Britannica* the fifteenth to nineteenth digits and presented them as a table of random numbers. Certainly the largest table of random numbers is the Rand Corporation publication of 1 million random digits [6].

Table A.1 is abstracted from Hald's tables. You will note that about one-tenth of the digits are zeroes, one-tenth are ones, one-tenth are twos, etc. You will also note that the digits are reasonably well mixed.

Quite often the objection is voiced that "numbers recorded in a table cannot possibly be random," or "the numbers cannot be random after they are written down." Two points should be emphasized.

1. A random number table is not random but is used **to simulate** or approximate the use of a random device.

2. We would **not** necessarily wish to record the results from a random device as random numbers. For example, a run of 1000 zeroes would be possible from the spinning of a wheel, but we would not want to record a run of 1000 zeroes in our table.

These ideas may require some thought in order to be accepted. How could digits recorded from a table of logarithms be more acceptable as a random number table than results obtained from a random device? The point is that in using a random number table we are more concerned with the output from the table than we are with the source of the table. We want the table **to simulate** the use of a random device, not record results from a random device.

4.7 We have discussed the canonical situation of drawing a random sample from an existent population. This provides a very useful way of thinking about data and, therefore, of making progress in science. Frequently, however, the sample is real but the population is only **conceptual**. We may have tossed a coin 50 times and obtained 30 heads. These 50 tosses represent a random sample from the conceptual population of all possible tosses. A similar situation also occurs frequently in the experimental sciences. Suppose that we wish to compare the pain-relieving effect of two new drugs, A and B. We select 50 subjects and assign A to 25 and B to the other 25. Our sample consists of these 50 subjects, but what is the population? The population might be a much larger group of people, one-half of whom are using A and one-half of whom are using B. But, in fact, no such population exists and is conceived of by the experimenter as a population that could exist for which our sample of 50 subjects could be a random sample.

4.8 One other departure from the canonical situation arises when the population may be real or conceptual but is **not well specified**. For example, suppose that an economist looks at price data for the last 70 years. She may choose to regard those 70 years' price data as a random sample from some population, necessarily vaguely specified (actually, such data are referred to as time series).

As you continue your study of statistics, difficulties will arise. Some of them can be surmounted if you insist on making the distinction between population and sample.

SUMMARY. Explicit examples of **sampling** from some **population** can be found in the early literature. However, recognition that only **samples** are being obtained from **populations** is often lacking. Although the distinction between sample and population was generally known by the end of the nineteenth century, the **distinction** was not always maintained.

In the early 1900s the distinction became clearer and was emphasized strongly by **Fisher** in 1922. Following Fisher, statistical theory and methods were developed to deal with the problem of making inferences about a **population** using a **random sample** from the population.

Implementation of a theory of random sampling requires a random device or a simulator of a random device. A very convenient simulator is provided by a random number table.

Given a **random sample** from a population, statistical theory is concerned with making **estimates** of **parameters**, calculating **intervals** of uncertainty, and performing **statistical tests.**

Finally, statistical theory deals with situations where the population is poorly defined or even **conceptual**.

REFERENCES

1. Carnap, Rudolph. (1962). *Logical Foundations of Probability*, Chicago: The University of Chicago Press.
2. Fisher, R. A. (1922). "On the Foundation of Theoretical Statistics," *Philosophical Transactions of the Royal Society, A-222*, 309–368.
3. Fisher, R. A., and Frank Yates. (1963). *Statistical Tables for Biological, Agricultural, and Medical Research*, 6th Edition, Edinburgh: Oliver & Boyd.
4. Hald, A. (1952). *Statistical Tables and Formulas*, New York: Wiley.
5. Peirce, C. S. (1958). *Collected Papers of Charles Sanders Peirce* edited by A. Burks, Cambridge, Mass.: Harvard University Press. The material quoted is from Lessons from the History of Science, written around 1896.
6. Rand Corporation. (1955). *A Million Random Digits with 100,000 Normal Deviates*, Glencoe, Ill.: Free Press.
7. Stephan, Frederick F. (1948). "History of the Uses of Modern Sampling Procedures," *Journal of the American Statistical Association, 43*, 12–39.

8. Tippett, L. H. C. (1927). *Random Sampling Numbers, Tracts for Computers*, Vol. 15, Cambridge, England: Cambridge University Press.
9. Yule, G. Udny. (1919). *An Introduction to the Theory of Statistics*, 5th Edition (1st Edition, 1910), London: Charles Griffin.

EXERCISES

1. Consider the population of sentences in Lecture 4 and suppose that you are concerned with the sentence length. How would you choose a random sample of 10 sentences?
2. Would you consider the students in this course to be a random sample of the students in the university? Why or why not?
3. Use Table A.1 to draw a random sample of two-digit numbers from a population with equal proportions of the numbers 00, 01, 02, . . . , 99. Choose your sample in the following way.
   a. Pick the starting point by placing a pencil "at random" on the page.
   b. Decide in advance whether to read horizontally, diagonally, vertically, etc.
   c. Read two-digit numbers (ignoring spacing in the table) starting at the starting point and using the rule decided on in part b.

   Does your sample seem reasonable?
4. Survey the statistics books in your library to find random number tables other than the ones mentioned in this section.
5. Not all who are eligible to do so vote in elections; those who vote constitute a sample of the population.
   a. Is it reasonable to consider the people who vote as a random sample of the population?
   b. Why are pre-election opinion polls highly suspect when there is low voter turnout?
6. Use Table A.1 to draw a random sample of 10 distinct three-digit numbers. Use these 10 numbers as page numbers of this book.
   a. In what ways do the sample pages resemble the book as a whole?
   b. In what ways are they markedly different?
   c. Use the sample pages to estimate the fraction of the book devoted to exercises.

7. The local weather forecast often gives the probability of rain, implying randomness of some sort. The local daily rainfalls for the past year are a sample of some population. What is the population?
8. Small opinion surveys are often reported on television with the statement that the results are not intended to be scientific—they are merely a random sampling of opinion. The first 10 people leaving an exclusive men's store are asked for their opinion.
   a. What is the population being sampled?
   b. Is the sample random?
   c. For what population could the sample be considered a random sample?

KARL PEARSON. 1857–1936. Reproduced from *Karl Pearson 1857–1957*, © 1958, Cambridge University Press, by permission from the *Biometrika* trustees.

5

FREQUENCY DISTRIBUTIONS AND THEORETICAL CURVES

5.1 The first volume of *Biometrika* [9] begins with an editorial containing the following statement.

*The first condition necessary, in order that any process of Natural Selection may begin among a race, or species, is the existence of differences among its members, and the first step in an inquiry into the possible effect of a selective process upon any character of a race must be an estimate of the frequency with which individuals, exhibiting any given degree of abnormality with respect to that character, occur.**

This emphasis on **frequency distributions** typifies the scientific literature of the period, which abounded with frequency distributions of all sorts of data.

5.2 In 1898 Bortkiewicz [1] gave the frequency distribution of the number of men per cavalry corps killed in 1 year by a kick from a horse (see Table 5.1). The early volumes of *Biometrika* contain many similar tabulations on things such as age of criminals, brain size, number of sepals on flowers, etc. Latter [4] gave the frequency distribution on the size of cuckoo eggs (Table 5.2).

Inspection of modern research journals will also reveal frequency distributions on data from widely varying sources. Maddi [5] gave the data in Table 5.3 on rating of attorneys by presiding judges.

Table 5.1 Number of Men Killed by Horsekicks (per Corps per Year)

Number Killed	Frequency
0	109
1	65
2	22
3	3
4	1

The data in Table 5.4 were given by Kjetsaa [3] in a study of the disputed authorship of *The Quiet Don.* The data give the number of different words used per 1000 words for three works—two by known authors and one of disputed authorship. They show that *The Quiet Don* resembles the work by Sholokhov more than the work by Kryukov. This is particularly interesting in view of the fact that Sholokhov was awarded the Nobel Prize for Literature in 1965 for *The Quiet Don.* Refer to the article by Kjetsaa for full details.

Table 5.2 Frequency Distribution of Cuckoo's Eggs

Breadth	Frequency (number)
13.75–14.25	1
14.25–14.75	1
14.75–15.25	5
15.25–15.75	9
15.75–16.25	73
16.25–16.75	51
16.75–17.25	80
17.25–17.75	15
17.75–18.25	7
18.25–18.75	0
18.75–19.25	1

Source. Reproduced from *Biometrika, 1,* Latter, The egg of cuculus canorus.

Table 5.3 Judges' Rating of Attorney Competence

Percent Attorneys Rated Competent	Judges Reporting Frequency
>99	4.0
90–99	13.3
80–89	24.4
70–79	31.7
60–69	5.8
50–59	7.5
40–49	6.5
30–39	3.5
20–29	1.9
10–19	1.3
1–9	0.1
<1	0.0

Source. Reproduced from "Trial Advocacy Competence: The Judicial Perspective," *Research Journal, 1978* (1), 46 pp.

The English clay tobacco pipe often provides valuable clues to students of historical sites. In 1954 Harrington published a chart relating the diameter of the hole in the stem to the date of the pipe. These data were given by Hume [2] and are shown in Table 5.5. From this table it can be seen that measurement of the hole diameter would be of value in dating a pipe fragment.

Table 5.4 Disputed Authorship—*The Quiet Don*

Work	Words Sampled	Distinct Words
Marking Time (Kryukov)	1000	589
The Way and the Road (Sholokhov)	1000	656
The Quiet Don	1000	646

Source. Reproduced from *Computers and the Humanities,* Vol. 11, Kjetsaa, The battle of the quiet don, Table I, p. 342.

Table 5.5 Stem Hole Diameter versus Date (Percentage)

Diameter (inches)	Date 1620–1650	1650–1680	1680–1710	1710–1750	1750–1790
$\frac{4}{64}$	0	0	0	13	77
$\frac{5}{64}$	0	0	12	72	20
$\frac{6}{64}$	0	18	72	15	3
$\frac{7}{64}$	21	57	16	0	0
$\frac{8}{64}$	59	25	0	0	0
$\frac{9}{64}$	20	0	0	0	0

Source. Reproduced from Figure 96, p. 298 of *Artifacts of Colonial America* by Ivor Noel Hume. Copyright © 1970 by Alfred A. Knopf, Inc. Reprinted by permission of Alfred A. Knopf. Inc.

5.3 Consider Table 5.2. Should an egg measuring exactly 14.75 be counted in the class 14.25–14.75 or the class 14.75–15.25? Some textbooks take a sophisticated approach to answering this question. It is unnecessary to make the matter complicated; it is necessary to make the matter clear. One might adopt the convention that the class 14.25–14.75 means up to but not including 14.75, the class 14.75–15.25 means up to but not including 15.25, etc. Whatever convention is adopted should be stated by the person presenting the data. Another way of resolving the question of **boundaries** between the **classes** is suggested by Table 5.2. If the breadth is measured to the nearest half-inch the classes have been selected so that it is impossible for a measurement to fall exactly on the boundary.

These questions do not arise in the data of Table 5.1 and Table 5.4. The measurements in these tables are examples of **discrete** measurements. No matter how carefully one counts, the only possible measurements are 0, 1, 2, 3, 4,.... By contrast, the measurements possible in Tables 5.2, 5.3, and 5.5 depend on how finely one measures, and they are called **continuous** measurements.

5.4 When we construct a frequency distribution for continuous data, grouping of the data into classes is required. Grouping may be done with discrete as well as continuous data, and the appearance of the frequency distribution depends on the number of classes.

Instead of tabulating the frequency of values in a class, we may record the **relative frequency** (the frequency divided by the total number of observations). Alternatively, we may record the percentage (the relative frequency multiplied by 100), as in Table 5.5.

5.5 It is often helpful to graph discrete frequency distributions as in Figure 5.1. Here the height of the dot represents the relative frequency obtained from Table 5.1. With continuous data it is often more helpful to graph the frequency distribution using a histogram. Here the relative frequency is represented by areas, and $f(x)$ is the relative frequency multiplied by the appropriate factor, so that the total area is one. This is illustrated in Figure 5.2 for the data of Table 5.2. The midpoint of each class is called the **class mark**; if one does calculations from the histogram, the class mark is used to represent all values belonging to that class.

5.6 In recent years the general public has become aware of photographs of the earth, the moon, Jupiter, Mars, and Venus that have been "**enhanced** by computer technology." Although the complete process is impressive, some of the basic concepts can be described in terms of the elementary concepts of this chapter. Scollar [8] gives the picture, prior to

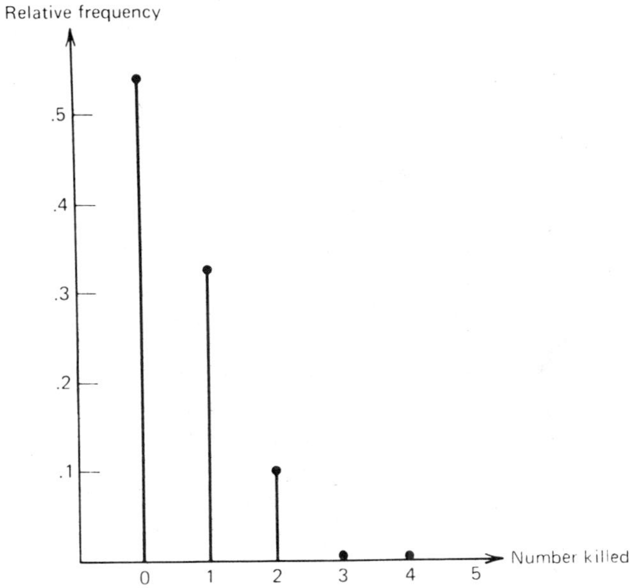

Figure 5.1 Frequency Distribution: Horsekick Data.

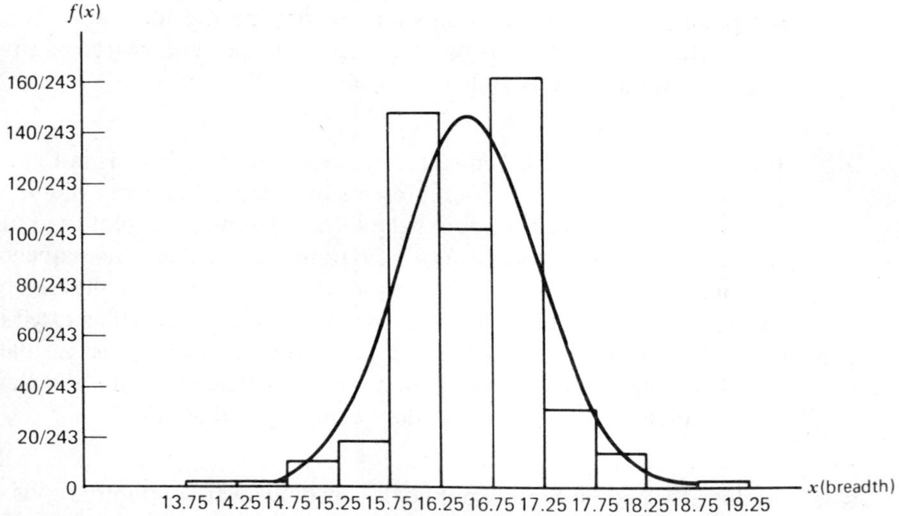

Figure 5.2 Frequency Distribution: Cuckoo Eggs.

enhancement, in Figure 5.3. By marking off the picture into small square areas and recording the shade of gray in each picture, one can build a histogram of shades of gray. This is shown in Figure 5.4. The various peaks represent the light and dark areas of the picture. Think about what can be done to the picture to **sharpen** the images. It seems quite reasonable to make the bright areas brighter and to darken the dark areas. This results in the **enhanced** picture shown in Figure 5.5.

The process of going from the original picture to the enhanced picture amounts to replacing a histogram with little spread by one with a great deal of spread. This is the same process that is involved when we turn the contrast knob on our television sets. We are **spreading out** the histogram.

5.7 As scientists constructed frequency distributions and graphs such as those presented in this lecture, similarities were noted, and this suggested representing distributions by mathematical equations.

Bortkiewicz noted that a probability formula published by Poisson [7] in 1837 gave numbers remarkably close to the observed relative frequencies. Using the Poisson formula

$$P(x) = \frac{(0.61)^x e^{-0.61}}{x!}$$

we can calculate the probability entries in Table 5.6.

Figure 5.3 Original Photograph. (Reproduced from *World Archaeology, 10,* Scollar, Computer image processing for archaeological air photographs, Plate 3. Copyright © 1978. Reprinted by permission of Rheinisches Landesmuseum, Bonn.) Courtesy of the American Museum of Natural History

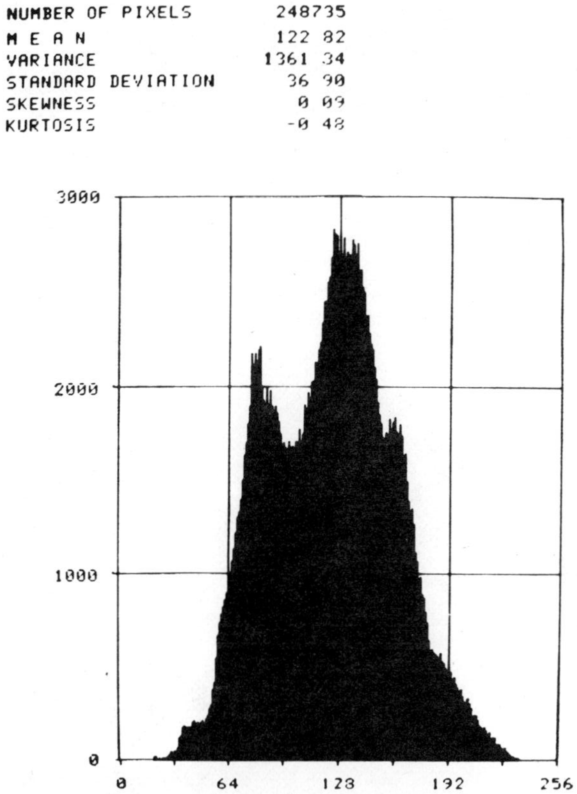

Figure 5.4 Frequency Distribution: Gray Values. (Reproduced from *World Archaeology, 10,* Scollar, Computer image processing for archaeological air photographs, Figure 3. Copyright © 1978. Reprinted by permission of Rheinisches Landesmuseum, Bonn.) Courtesy of the American Museum of Natural History

Figure 5.5 Enhanced Photograph. (Reproduced from *World Archaeology, 10,* Scollar, Computer image processing for archaeological air photographs, Plate 4. Copyright © 1978. Reprinted by permission of Rheinisches Landesmuseum, Bonn.)

Table 5.6 Relative Frequency Distribution

Number Killed	Relative Frequency	Probability
0	0.545	0.544
1	0.325	0.331
2	0.110	0.101
3	0.015	0.021
4	0.005	0.003

For $x = 0$ we get 0.544, for $x = 1$ we get 0.331, etc. In a later lecture we will say more about the Poisson distribution and explain how 0.61 was obtained. For the moment, consider the implication of having a mathematical formula. This suggests the possibility of finding a generating mechanism—a scientific law.

The probabilities are also shown in Figure 5.1.

5.8 The Gaussian probability curve discussed in Lecture 3 was discovered to be useful for describing frequency distributions from many varied sources. Because of its wide applicability, Pearson called it **the normal distribution** in a Gresham College lecture in 1893. Consider what would happen to the histogram in Figure 5.2 if measurements were refined. The classes become smaller and smaller, and the boundary of the histogram approaches a smooth, bell-shaped curve with equation

$$f(x) = \frac{1}{\sigma\sqrt{2\pi}} e^{-(x-\mu)^2/2\sigma^2}$$

5.9 The consequences of representing population frequency distributions by regular mathematical expressions have been great. The power of probability theory has been brought to bear on the development of statistical theory. Consequently, statistical methodology rests not on ad hoc methods but on rigorous statistical theory. A transition in thinking about probability models has taken place among theoretically trained statisticians. In the beginning mathematical models were used slowly, cautiously, as approximations to the frequency distributions of real populations. Today some people regard the probability models as reality and more irregular frequency distributions as approximations to reality. In my judgment, we should continue to regard probability models as approximate descriptions, sometimes remarkably good, of frequency distributions in the "real" world.

SUMMARY. The scientific literature of the late 1800s and early 1900s abounded with **frequency distributions** of all sorts of data, as did the early volumes of *Biometrika*. This lecture presents frequency distributions on the number of soldiers killed by **horsekicks**, breadth of **cuckoo eggs**, ratings of **attorney competence**, frequency of words used in a work of **disputed authorship**, diameters of **clay pipe stems**, and shades of gray in **image enhancement**.

The grouping of data into **classes** is discussed. This results in **class boundaries** and **class marks**, the midpoints of each class. For each class the **frequency** or **relative frequency** of observations is recorded.

With **continuous** observations the choice of classes affects the frequency distribution, no matter how finely one measures. With **discrete** observations the possible values are the same, regardless of how fine the measurement.

In many cases the data can be well described by a theoretical **probability formula**. Two examples are given: the **Poisson** formula for the horsekick data and the **normal distribution** for the egg data.

REFERENCES

1. Bortkiewicz, L. (1898). *Das Gesetz der kleinen Zahlen*, Leipzig: Teubner.
2. Hume, Ivor Noël. (1970). *A Guide to Colonial Artifacts*, New York: Knopf.
3. Kjetsaa, Geir. (1977). The Battle of The Quiet Don," *Computers and the Humanities, 11*, 341–346.
4. Latter, Oswald H. (1902). "The Egg of Cuculus Canorus," *Biometrika, 1*, 164–176.
5. Maddi, Dorothy Lender. (1978). "Trial Advocacy Competence: The Judicial Perspective," *American Bar Foundation Research Journal, 1978*, 105–151.
6. Mosteller, F., and D. L. Wallace. (1964). *Inference and Disputed Authorship: The Federalist*, Reading, Mass.: Addison-Wesley.
7. Poisson, S. D. (1837). *Recherches sur la probabilité des jugements en matière criminelle et en matière civile*, Paris.
8. Scollar, Irwin. (1978). "Computer Image Processing for Archaeological Air Photographs," *World Archaeology, 10*, 71–87.
9. Weldon, W. F. R., and Karl Pearson. (1901). Editorial, *Biometrika, 1*, 1–2.

EXERCISES

1. At a small college the undergraduate English majors intending to attend graduate school took the Graduate Record Examination. The verbal scores were 552, 278, 342, 675, 725, 340, 482, 495, 627, 539, 428, 740, 329, 370, 590, and 462.

a. Using classes 200–300, 300–400, ..., construct the frequency distribution. State the convention you use concerning scores that fall on the class boundary.
b. What class boundaries could you use that would give the same frequency distribution but would avoid the problem of observations falling on the boundaries?

2. An owner of a large, 1975 automobile becomes interested in trading for a used, later-model small car and begins to study the used car lots. On a large lot of 40 cars, he jots down the ages of the cars on the lot. The ages to the nearest year are 3, 7, 3, 4, 5, 1, 2, 3, 4, 5, 9, 5, 6, 5, 4, 8, 4, 5, 2, 1, 6, 5, 6, 3, 2, 1, 7, 4, 5, 5, 4, 3, 2, 3, 4, 5, 8, 6, 6, and 5.
   a. Construct the frequency distribution.
   b. Graph the frequency distribution obtained in part a, letting the ordinate value represent frequency.
   c. Using classes 0.5–1.5, 1.5–2.5, etc., represent the frequency distribution by a histogram so that the areas represent frequencies.
3. In making a study of office operations, the manager becomes concerned with the amount of time spent processing incoming junk mail. She asks the administrative secretary to keep a record of incoming mail. The daily number of pieces of junk mail for a 4-week period were 55, 127, 89, 72, 84, 77, 115, 130, 116, 120, 140, 125, 131, 115, 99, 126, 145, 124, 89, and 78.
   a. Construct the frequency distribution using the cells 54.5–74.5, 74.5–94.5, 94.5–114.5, 114.5–134.5, and 134.5–154.5.
   b. Represent the frequency distribution by a histogram and label the vertical axis so that frequencies are represented by areas.
4. A local car dealer advertises actual gasoline mileage in addition to EPA estimates. A group of 24 people who have purchased cars of the same model decide to compare their experiences with the advertising. The mileages (miles per gallon) were 18.7, 21.2, 17.8, 15.9, 16.0, 19.0, 18.6, 18.8, 20.2, 19.4, 19.3, 18.7, 16.1, 17.2, 19.2, 18.7, 19.3, 19.9, 18.8, 18.1, 17.7, 20.2, and 20.3.
   a. Using classes 15.05–16.05, 16.05–17.05, etc., construct the frequency distribution and histogram.
   b. Does the histogram seem to support the advertised claim of 19 miles per gallon?
5. Construct a frequency distribution for the sample obtained in Exercise 3 of the previous lecture. Use classes 00–09, 10–19, ..., 90–99. Does the frequency distribution seem reasonable for a random number table?

6. Using a current road atlas, construct a frequency distribution for the city sizes in your state. You will need to choose classes so that the number of classes is not excessive.
7. Using the newspaper file in your library, construct a frequency distribution for the New York Stock Exchange closing prices for a stock of your choice over the past 3 months.
8. Construct a frequency distribution for the coat sizes found in a local clothing store. Note that the clothing rack itself gives a visual presentation of a frequency distribution.
9. Mosteller and Wallace [6] use frequency distributions of basic words as a way of studying papers of disputed authorship. Construct a frequency distribution for the number of occurrences of the word "of" per 100 words in this book. Do the same for some other statistics book and compare the distributions. You need only examine a few pages in each book.
10. Using classes 13.75–14.75, 14.75–15.75, ... for the data of Table 5.2, construct a frequency distribution and a histogram. How does the histogram compare with the histogram of Figure 5.2?
11. From the data in Table 5.3, what percent of the the judges rated at least 70% of the attorneys competent? At least 50% of the attorneys competent?
12. Construct a histogram of clay pipe stem hole diameters for each of the time periods given in Table 5.5.
13. Construct your own image enhancement project. Shade a simple picture using only three shades of gray (possibly by using pencils of different hardness). Divide the picture, using horizontal and vertical lines, into 100 small, rectangular regions. Now build a frequency distribution and histogram for the three shades of gray. Do a second version of the picture involving five shades of gray, using lighter shades for the lights and darker shades for the darks. Construct the frequency distribution and histogram for the enhanced picture.

TABLES, ſhewing the Probabilities of the Duration of Human Life, from the Birth up to ninety years of age---for divers intermediate Periods of Life;---at the City of Philadelphia, and at the Town of Salem in Maſſachuſetts; and alſo in ſeveral parts of Europe.——

GENERAL Table of the Probabilities of Life, from the Calculations of *M. Buffon*.

Periods of Life.	Perſons living.	Decreaſe of Life.
	1,000	——
Between the Birth and 3	591	409
3 and 5	540	51
5 and 10	490	50
10 and 20	450	40
20 and 30	392	58
30 and 40	323	69
40 and 50	252	71
50 and 60	180	72
60 and 70	101	79
70 and 80	27.63	73.37
80 and 90	3.54	24.09

N°. 1. PHILADELPHIA, for twenty-two Years.

Periods of Life.	Perſons living.	Decreaſe of Life.
	1,000	——
Between the Birth and 3	612	388
3 and 5	555	57
5 and 10	511	44
10 and 20	465	46
20 and 30	368	97
30 and 40	270	98
40 and 50	178	92
50 and 60	114	64
60 and 70	52	62
70 and 80	20	32
80 and 90	5.61	14.39

N°. 2. *(a)* PHILADELPHIA, for 1782, 1788, 1789 and 1790.

Periods of Life.	Perſons living.	Decreaſe of Life.
	1,000	——
Between the Birth and 3	611	389
3 and 5	569	42
5 and 10	546	23
10 and 20	497	49
20 and 30	400	97
30 and 40	296	104
40 and 50	195	101
50 and 60	140	55
60 and 70	62	78
70 and 80	25	37
80 and 90	6	19

SALEM *(in Maſſachuſetts)* for 1782, *(b)* 1783, 1789 and 1790.

Periods of Life.	Perſons living.	Decreaſe of Life.
	1,000	——
Between the Birth and 3	----	——
3 and 5	555	445
5 and 10	505	50
10 and 20	470	35
20 and 30	342	128
30 and 40	252	90
40 and 50	169	83
50 and 60	129	40
60 and 70	94	35
70 and 80	26	68
80 and 90	*(c)*	

SALEM *(in Maſſachuſetts,) for* 1790. *(d)*

Periods of Life.	Perſons living.	Decreaſe of Life.
	1,000	——
Between the Birth and 3	----	----
3 and 5	550	450
5 and 10	503	47
10 and 20	487	16
20 and 30	356	131
30 and 40	293	63
40 and 50	220	73
50 and 60	178	42
60 and 70	120	52
70 and 80	42	84
80 and 90	15.71	26.29

PARIS, From the Calculations of M. Buffon. *(e)*

Periods of Life.	Perſons living.	Decreaſe of Life.
	1,000	——
Between the Birth and 3	----	----
3 and 5	580	420
5 and 10	524	56
10 and 20	485	39
20 and 30	433	52
30 and 40	366	67
40 and 50	293	73
50 and 60	212	81
60 and 70	116	96
70 and 80	32	84
80 and 90	4.50	27.50

MORTALITY TABLES. Births and deaths in preindustrial times. *Mortality in Preindustrial Times*, Gregg International Publishers, 1973. From *Observations on the Probabilities of the Duration of Human Life and the Progress of Population in the United States, 1793*, in a letter from William Barton to David Rittenhouse.

6

MOMENTS AND PERCENTILES

6.1 *During the same winter the Plateans, who were still being besieged by the Peloponnesians and the Boeotians, began to be distressed by failure of their supply of food, and since there was no hope of aid from Athens nor any other means of safety in sight, they and the Athenians who were besieged with them planned to leave the city and climb over the enemy's walls, in the hope that they might be able to force a passage.... They made ladders equal in height to the enemy's wall, getting the measure by counting the layers of bricks at a point where the enemy's wall on the side facing Platea happened not to have been plastered over. Many counted the layers at the same time, and while some were sure to make a mistake, the majority were likely to hit the true count, especially since they counted time and again and, besides, were at no great distance, and the part of the wall they wished to see was easily visible. The measurement of the ladders then, they got at in this way, reckoning the measure from the thickness of the bricks** [4].

This example of the use of the **mode**, or most popular value, in the fifth century B.C. was cited by Wallis and Roberts [6]. The practice of forming some sort of central value for data is certainly very old; Walker [5] traces the use of the **average** to the time of Pythagoras. The concept of a central value in data is called **central tendency** in statistics, and there are many measures of the concept; the most usual measures are the **mean**, **median**, and **mode**. The mean is the average value, the median is the central value, and the mode is the most popular value. With the numbers 1, 2, 2, 3, 4, 7, 9, the three quantities are given by:

$$\begin{aligned} \text{Mean} &= 4 \\ \text{Median} &= 3 \\ \text{Mode} &= 2 \end{aligned}$$

All three measures are valuable, but the mean is the most widely used.

6.2 Most schoolchildren are usually taught to calculate the arithmetic average, or mean, in the United States during the fourth or fifth grade. We wish to relate the elementary definition of the mean to the frequency distributions of previous lectures and to the probability distributions of the lectures to follow. Therefore we will examine the arithmetic mean in minute detail and rewrite it in a way that makes its relation to frequency distributions more obvious. Consider a population consisting of the numbers 2, 3, 2, 4, 2, 3, 4, 5, 3, 2. Then the population mean, μ, is given by

$$\begin{aligned} \mu &= \frac{2+3+2+4+2+3+4+5+3+2}{10} \\ &= 3 \end{aligned}$$

Regrouping the numbers in the numerator, we obtain the alternative forms

$$\begin{aligned} \mu &= \frac{(2)(4)+(3)(3)+(4)(2)+(5)(1)}{10} \\ &= 2(4/10)+3(3/10)+4(2/10)+5(1/10) \end{aligned}$$

We are therefore led to the simple formula:

$$\mu = \sum_{i=1}^{k} x_i f_i$$

where x_i denotes the distinct values, f_i denotes the relative frequency of distinct values, and k denotes the number of distinct values. The calculation may be presented in tabular form, as in Table 6.1.

If these same calculations are carried out on a sample, we denote the sample mean by $\bar{x}$ and regard it as an *estimate* of the population mean, μ.

If an individual is selected at random from the small population of the

Table 6.1 Calculation of Mean

x_i	f_i	$x_i f_i$
2	$\frac{4}{10}$	$\frac{8}{10}$
3	$\frac{3}{10}$	$\frac{9}{10}$
4	$\frac{2}{10}$	$\frac{8}{10}$
5	$\frac{1}{10}$	$\frac{5}{10}$
Total	1	$\mu = 3$

preceding example, the relative frequencies become probabilities and the formula for the **population mean** becomes

$$\mu = \sum_{i=1}^{k} x_i P(x_i)$$

6.3 By the time of Pascal and Fermat in the seventeenth century, a concept more primitive than probability, that of **expectation**, was already well established. In fact, in early papers probability was defined in terms of expected earnings. Life expectancy has also been a familiar concept for several hundred years since the appearance of the first mortality tables for human life. We now want to relate expectation (expectancy, expected value) to the mean. Consider Table 6.2.

Although the table does not contain the information necessary to verify its entries, we should recognize the population quantities it estimates

Table 6.2 Life Expectancy (Years)

Age	1900–1902	1976
0	49.24	72.8
10	51.14	64.2
20	42.79	54.6
30	35.51	45.3
40	28.34	35.9
50	21.26	27.2
60	14.76	19.4
70	9.30	12.9
80	5.30	7.9

Source. National Center for Health Statistics.

and the sort of data that would be required to verify the estimates. For example, the entry of 49.24 is an estimate of the average lifetime for all of the people born in the United States between 1900 and 1902.

Conceptually, it would be possible to calculate this number by the methods of the previous section once all of the lifetimes have been determined and recorded. By multiplying each lifetime by its relative frequency and summing, the mean lifetime could be determined. It is called the **life expectancy** at age zero, or at birth. Similarly, the mean remaining lifetime could be determined for all the people of age 30 born between 1900 and 1902 or of age 20 born in 1976, etc. The life expectancy at any age is simply the mean remaining lifetime for all people of that age. The figures in Table 6.2 are estimates based on probability models and extensive data.

Expected value has come to be synonymous with mean value; the mean value of X is often denoted by $E(X)$ and read as "the expected value of X." Since the mean value is not always a member of the population, one should be reminded that the **expected value** is not the value we expect; it is the average or mean value. To summarize,

$$\mu = E(X) = \sum_{i=1}^{k} x_i f_i = \sum_{i=1}^{k} x_i P(x_i)$$

Let us repeat the steps of Section 6.2 in order to calculate the expected value of X^2. Then

$$E(X^2) = \frac{2^2 + 3^2 + 2^2 + 4^2 + 2^2 + 3^2 + 4^2 + 5^2 + 3^2 + 2^2}{10}$$

$$= 2^2(4/10) + 3^2(3/10) + 4^2(2/10) + 5^2(1/10)$$

We are then led to the formula

$$E(X^2) = \sum_{i=1}^{k} x_i^2 f_i = \sum_{i=1}^{k} x_i^2 P(x_i)$$

We easily conclude that

$$E(X^3) = \sum_{i=1}^{k} x_i^3 f_i = \sum_{i=1}^{k} x_i^3 P(x_i)$$

and, in fact for any function $g(X)$ of X,

$$E[g(X)] = \sum_{i=1}^{k} g(x_i) f_i = \sum_{i=1}^{k} g(x_i) P(x_i)$$

Follow the details of the foregoing development carefully, because doing so will provide an understanding of formulas that are meant to convey simple ideas but that can look formidable.

6.4 In the preceding sections we have introduced the concept of **central tendency** and several measures of that concept, with primary emphasis on the mean or expected value. In this section we want to consider the concept of **variation**, or dispersion, and measures of that concept. By the late 1800s well-defined measures of variation were in use. Pearson [3] is credited with the name **standard deviation** for the measure defined as the square root of the mean of the squares of deviations from the mean of the distribution. The symbol σ was used to denote the standard deviation. In terms of the previous section,

$$\sigma = \sqrt{E[(X-\mu)^2]}$$
$$= \sqrt{\sum_{i=1}^{k} (x_i - \mu)^2 f_i}$$

For the example of Table 6.1 we illustrate the calculation of the standard deviation in Table 6.3.

A word of caution may be helpful at this point. Beginning students may allow a few symbols to prevent them from trying to grasp the main ideas. Without practice one may forget how to calculate σ. For the nonstatistician a general knowledge of the noncomputational concepts may be more viable in the long run. In all populations of numbers there is **variation**. There are many **measures** of variation; one of the most useful measures is the standard deviation. If there is no variation, if all of the numbers are equal, the standard deviation is zero. Otherwise it is positive.

Table 6.3 Calculation of Standard Deviation

x_i	f_i	$(x_i-\mu)^2$	$(x_i-\mu)^2 f_i$
2	$\frac{4}{10}$	1	$\frac{4}{10}$
3	$\frac{3}{10}$	0	0
4	$\frac{2}{10}$	1	$\frac{2}{10}$
5	$\frac{1}{10}$	4	$\frac{4}{10}$
Total	1		$\sigma^2 = 1$ $\sigma = 1$

In 1918 Fisher [2] introduced the name **variance** to denote σ^2, the square of the standard deviation. It has virtually the same properties as the standard deviation but has proven to be easier to work with in many situations.

If we do the calculations in Table 6.3 with a sample instead of a population, we call the sample variance an **estimate** and denote it by $\hat{\sigma}^2$ or frequently by $s^2 = \Sigma(x_i - \bar{x})^2/(n - 1)$.

6.5 The analogy between the **moments** of mechanics and the averages of statistics was noticed by early workers in the field of probability, and Pearson apparently first used the word moment in a statistical sense in 1893 (see Walker [5, p. 73]). It may seem startling that there is any relationship between life expectancy and the **center of gravity** of a bridge structure but, once the two quantities are formulated mathematically, the relationship is obvious.

Just as statistical averaging goes back to antiquity, the use of moments in the study of mechanics can be traced to the ancient writers. Cajori [1] credits Archimedes with the theory of the center of gravity, called the first moment.

Suppose that k objects of unequal mass are placed on a board and balanced on a fulcrum, as in Figure 6.1. The location of the center of gravity is easily obtained from the formula

$$c = \frac{\sum_{i=1}^{k} x_i M_i}{\Sigma M_i}$$

where

x_i = distance of ith object from the left end of the board
M_i = mass of the ith object

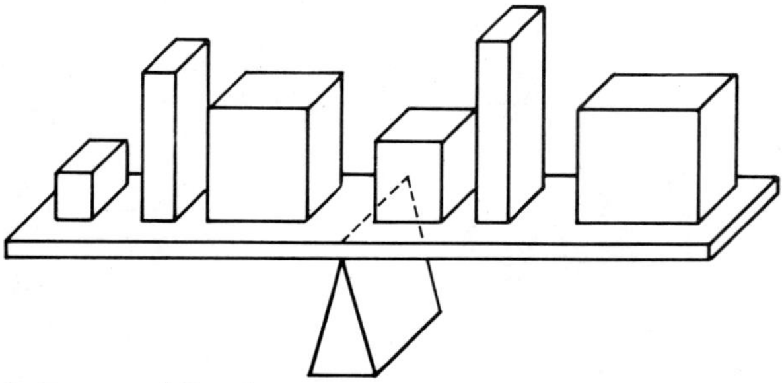

Figure 6.1 Center of Gravity.

Now, if we denote the relative mass by $m_i = M_i/\Sigma M_i$, the formula becomes

$$c = \sum_{i=1}^{k} x_i m_i$$

an expression completely equivalent to the mean or expected value.

In the construction of structures it is important to have some measure of the rotational inertia. A measure in use for hundreds of years is the **moment of inertia**, given by the formula

$$\text{Moment of inertia} = \sum_{i=1}^{k} (x_i - c)^2 M_i$$

A related measure is the **radius of gyration**, given by

$$(\text{Radius of gyration})^2 = \sum_{i=1}^{k} (x_i - c)^2 m_i$$

The analogy between mechanics and statistics can be made stronger by considering the graph of the frequency distribution. Graphing the frequency distribution of Section 6.2, we have Figure 6.2.

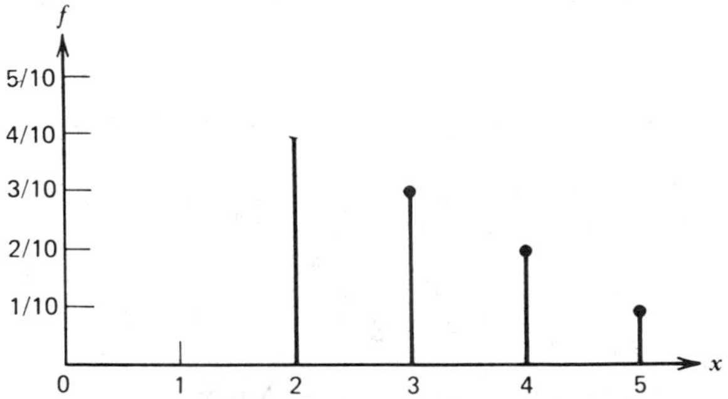

Figure 6.2 Discussion of Moments.

If we think of the probability as mass, the familiar mean, or average, can be thought of as the center of gravity of our frequency distribution. Similarly, the standard deviation can be thought of as the radius of gyration. This analogy has proven helpful to many people. If the numbers in the population are scattered widely about the mean, the distribution might be thought of as highly unstable with a large radius of gyration, or standard deviation.

6.6 Walker considers Galton to have been the originator of the terms **median** and **percentile**, although median was actually suggested earlier by Gauss. The basic idea of the median is that of dividing the population into two equal parts—large numbers and small numbers—and the median is the dividing line. There is some indeterminacy with the idea (e.g., consider the population 1, 1, 1, 1, 1, 2 for which there is no natural division) but, for populations of real data, there is little difficulty, and the median is used widely.

Percentiles were developed by Galton (see Walker) for purposes closely related to their modern usage. The basic idea is that we rank the numbers of the population from low to high and then divide the population into 100 equal parts. Then the ninetieth percentile is a number such that 90% of the numbers in the population are less and 10% are greater. It may be helpful to note that the fiftieth percentile is, by definition, the median.

6.7 Given only a few numbers, it is doubtful that one should perform any calculations. However, frequency distributions and graphs may help one to comprehend a large population. Furthermore, parameters such as the mean, standard deviation, and percentiles may be of assistance. However, it is only with experience that they can be of any great assistance in understanding a population of numbers.

SUMMARY. Measures of the central value can be traced at least as far as the fifth and sixth century B.C. Of the many measures possible, the most widely used are the **mean**, **median**, and **mode**. The mode is the most frequent value, the median is the middle value, and the mean is the average value.

The **expected value** of a function of X, $g(X)$, is defined to be $\Sigma g(x)P(x)$ and is denoted by $E[g(X)]$. This terminology is consistent with the use of the term **life expectancy** to describe the average lifetime of humans. With the definition of expectation, the **population mean** is simply the expected value of X, $E(X)$, and is denoted by μ. The **sample mean**, denoted by $\bar{X}$, is an **estimate** of the population mean.

The most usual measure of population **variation** is the **variance**, defined by $E[(X - \mu)^2]$ and denoted by σ^2. The square root of the variance, denoted by σ, is called the **standard deviation**. The sample variance and standard deviation are denoted by s^2 and s, respectively, and are used as estimates of the population variance and standard deviation.

The analogy between the mean and variance and the center of gravity and moment of inertia prompts the use of the word **moments** to describe the mean and variance.

In addition to the mean and variance, **percentiles** are also used as measures of **central tendency** and **variation**.

REFERENCES

1. Cajori, Florian. (1899). *A History of Physics*, New York: Macmillan.
2. Fisher, R. A. (1918). "The Correlation Between Relations on the Supposition of Mendelian Inheritance," *Transactions of the Royal Society of Edinburgh*, *52*, 399–433.
3. Pearson, Karl. (1894). Contributions to the Mathematical Theory of Evolution-I. On the Dissection of Asymmetrical Frequency Curves," *Philosophical Transactions*, *A-185*, 71–110.
4. Smith, Charles Forster. (1920). *History of the Peloponnesian War*, Books III and IV (English translation of Thucydides II), Cambridge, Mass.: Harvard University Press.
5. Walker, Helen M. (1929). *Studies in the History of Statistical Method*, Baltimore: Williams & Wilkins.
6. Wallis, W. A., and H. V. Roberts. (1956). *Statistics: A New Approach*, Glencoe, Ill.: Free Press.

EXERCISES

1. For the population of numbers 8, 8, 8, 8, 8, 8, 8, 8, 8, and 8, calculate μ, σ^2, the mode, and the median.
2. Show algebraically that the mean of a constant is the constant and that the standard deviation is zero.
3. For the two sets of numbers 10, 8, 7, 9, 5, 12, 8, 6, 8, 2 and 20, 16, 14, 18, 10, 24, 16, 12, 16, 4, calculate μ and σ^2.
4. Show algebraically that if the mean and variance of a set of numbers are μ and σ^2, the mean and variance of a set of numbers obtained by multiplying each original number by c are $c\mu$ and $c^2\sigma^2$.

5. Which of the following sets of data is more variable? Justify your answer.

Set 1	Set 2
10	110
18	118
25	125
16	116
22	122
30	130
26	126
24	124
25	125
23	123

6. Show algebraically that the mean and variance of a set of numbers obtained by adding a constant c to each of the original numbers are $\mu + c$ and σ^2.
7. What is the expected value, μ, of a one-digit number generated by a random device that assigns equal probability to the integers 0, 1, 2, . . . , 9?
8. What is the expected coat size, μ, for the frequency distribution of Lecture 5, Exercise 8? Do you think this is the expected coat size for the potential customers of the store?
9. Using the data of Table 5.1, estimate μ, the expected number of men killed by horsekicks. Is the number used in the equation given in Section 5.6? Estimate the standard deviation.
10. Using the data from Table 5.2, estimate the expected breadth of an egg, μ, and the standard deviation, σ. Do your calculation with the class marks 14.00, 14.50, ...
11. Explain the effect of the brightness and contrast controls on your television set in terms of changing the mean and standard deviation.
12. Verify by example that $E(X^3)$ is given by

$$E(X^3) = \sum_{i=1}^{k} x_i^{\,3} f_i$$

13. Repeat Exercise 10 on a coded set of values obtained by subtracting 16.00 from each class mark and then multiplying by 100. Denote the values obtained by $\bar{x}'$ and $s^{2'}$. What decoding must be done to obtain the values of $\bar{x}$ and s^2 in Exercise 10?

14. Verify algebraically the results implied by Exercise 13; i.e., if

$X' = a + bX,$

$$\mu' = a + b\mu$$

and

$$\sigma^{2'} = b^2\sigma^2$$

Note that this result describes image enhancement in that the mean of the histogram is shifted and the variance is changed.

15. Calculate the mean μ and variance σ^2 for the GRE scores of Lecture 5, Exercise 1. Also calculate the mean and variance using the class marks and frequencies of the frequency distribution.
16. Calculate the mean and variance for the age of automobiles on the used car lot described in Lecture 5, Exercise 2. Perform the same calculation using the class marks and frequencies of the frequency distribution. Are the results different? Why or why not?
17. Consider the data of Lecture 5, Exercise 3, to be a random sample. Calculate the sample mean and variance of the number of pieces of junk mail per day. Repeat the calculation using the class marks and frequencies of the frequency distribution.
18. Considering the 24 cases of Lecture 5, Exercise 4 as a sample, calculate the sample mean, $\bar{X}$, and variance, s^2, of gasoline mileage. Repeat the calculation using the class marks and frequencies of the frequency distribution.

PROBABILITY

PIERRE DE FERMAT. 1601–1665. Smith Collection, Rare Book and Manuscript Library, Columbia University.

7

INTERPRETATION OF PROBABILITY

7.1 Probability has many meanings. Those who use the English language are accustomed to words with multiple definitions, where the meaning depends on the context. However, it sometimes comes as a shock that words used in technical theories are not uniquely defined. How can words such as electron, mass, energy, and force have different meanings? How can probability have different interpretations?

The word probability is used basically in two different ways: (1) an **intrinsic property** of some system that does not depend on our knowledge of that system, and (2) a **measure of belief** in some statement. For years a controversy has existed among some probabilists and statisticians about the proper meaning of the word probability. At other times bitter controversy has resulted from a failure to distinguish between the two basic meanings. The argument goes back at least to von Mises [8] in 1919 and Keynes [4] in 1921 and continues to the present. The situation is illustrated by two books in elementary statistics. Savage [7] begins his book with the view that probability is a quantitative measure of uncertainty. Noether [5] presents the degree of belief interpretation as a secondary interpretation. The different uses of the word probability lie at the heart of the Bayesian controversy, which we discuss in Lecture 31.

7.2 Scientific theories are built on **undefined** (or ill-defined) terms. Attempts to define mass, force, and acceleration are unsatisfactory. Yet airplanes fly, locomotives move, and satellites orbit the earth because of theories based on these terms. Electrons have sometimes been described as particles, sometimes as waves, sometimes as both, and transistor technology advances even though the word electron is not well defined.

An analogous situation exists with probability. Statistical methods and probability models have proven themselves useful, although probability itself has **not** been well defined.

7.3 It is useful to distinguish between the concepts we are trying to describe and our attempted descriptions (models). Table 7.1 emphasizes this distinction.

If a lecturer asks a class for the probability that a regular coin will land heads, the answer will probably be a resounding "one-half." (Note that the previous sentence uses probability both as a **property** of the coin and as a **measure** of degree of belief.) The answer of one-half is undoubtedly a conditioned response; nevertheless, it is clear that the answer is an attempt to describe a **property of the coin** and not a **degree of belief**. The property is the long-run proportion of heads, or long-run relative frequency. To illustrate the idea, I tossed a coin 50 times. The results are shown in Table 7.2. The number of the toss is shown in the first column and is denoted by n. The result of the toss is shown in the second column under H or T. The total number of heads at any given toss is shown in the third column and is denoted by $n(H)$. Finally, the **relative frequency** of heads, calculated by $n(H)/n$, is given in the fourth column. These results are shown graphically in Figure 7.1. Note that after 50 tosses the relative frequency has settled down in the neighborhood of one-half. We would suspect that if we tossed further, the relative frequency would fluctuate above and below one-half but would settle even closer to the long-run relative frequency of one-half.

There are several coin-tossing experiments recorded in the literature. A well-known example is given by Kerrick [3]. While interned at Hald, Denmark, during World War II, he performed several experiments. For example, he tossed a coin 10,000 times and made 5000 draws of two balls from an urn. Initially in the coin-tossing experiment the relative frequency of heads fluctuated widely but, after 10,000 tosses, it settled down in the vicinity of one-half, with a value of 0.507 on the final toss.

Table 7.1 Interpretations and Models of Probability

Interpretation	Model
Property of a system (long-run frequency)	1. Mathematical limit 2. Collectives of von Mises
Measurement of degree of belief	1. Logical relation theory 2. Subjective probability

Table 7.2 Relative Frequency of Heads

n	H or T	$n(H)$	$n(H)/n$	n	H or T	$n(H)$	$n(H)/n$
1	H	1	1.0000	26	T	12	0.4615
2	H	2	1.0000	27	T	12	0.4444
3	H	3	1.0000	28	H	13	0.4643
4	T	3	0.7500	29	T	13	0.4483
5	H	4	0.8000	30	H	14	0.4667
6	H	5	0.8333	31	H	15	0.4839
7	H	6	0.8571	32	H	16	0.5000
8	T	6	0.7500	33	T	16	0.4848
9	T	6	0.6667	34	T	16	0.4706
10	H	7	0.7000	35	T	16	0.4571
11	H	8	0.7273	36	H	17	0.4722
12	H	9	0.7500	37	T	17	0.4595
13	T	9	0.6923	38	T	17	0.4474
14	H	10	0.7143	39	T	17	0.4359
15	T	10	0.6667	40	T	17	0.4250
16	T	10	0.6250	41	H	18	0.4390
17	T	10	0.5882	42	T	18	0.4286
18	T	10	0.5556	43	T	18	0.4186
19	T	10	0.5263	44	T	18	0.4091
20	T	10	0.5000	45	H	19	0.4222
21	H	11	0.5238	46	H	20	0.4348
22	T	11	0.5000	47	H	21	0.4468
23	T	11	0.4783	48	H	22	0.4583
24	H	12	0.5000	49	T	22	0.4490
25	T	12	0.4800	50	H	23	0.4600

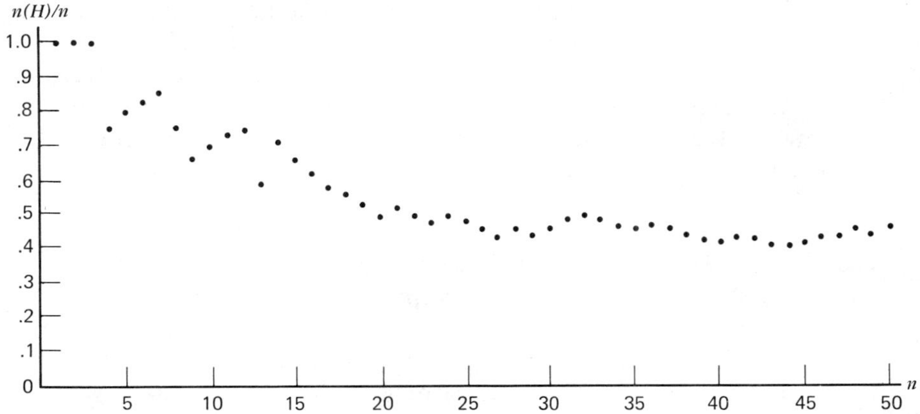

Figure 7.1 Relative Frequency of Heads.

The long-run relative frequency is not well-defined, but there is something real about it that any of us can experience by tossing a coin.

7.4 One of the most obvious ways to try to describe the long-run frequency property is in terms of a mathematical limit. This is often called the **limiting frequency definition** of probability. If n denotes the number of trials and $n(H)$ the number of occurrences of an event H, then the probability of H, denoted by $P(H)$, is defined by

$$P(H) = \lim_{n \to \infty} \frac{n(H)}{n}$$

This definition has proved to be very helpful, although it falls short of describing the phenomenon exhibited by coin tossing. If the relative frequency of heads approaches one-half as a mathematical limit, there would have to be a point where it was close to one-half and would never be further away with additional tosses. But we have to admit the possibility of an unlimited number of tails at any point. Thus the relative frequency of heads may be closer to one-half after 10,000 tosses than after 20,000 tosses. This is not allowed by a mathematical limit.

7.5 A vigorous effort to define probability by limiting frequency was made by von Mises. In order to avoid the difficulty of a sequence (such as the coin-tossing example) with no limiting frequency, von Mises restricted the definition of probability to regular sequences with limiting frequencies. He gave the name **collective (kollectiv)** to such a sequence. It is not generally felt that the work of von Mises resolved the difficulty inherent in the limiting frequency definition. As Hacking [1] states, "If there are skeptics who insist that the frequency in the long run with which the coin falls heads is no property of anything, they have this much right on their side: the property has never been clearly defined."

7.6 There are basically two theories of probability as a measure of belief: **logical relation** theory and **subjective** probability. Probability, as presented by the logical relation theory, measures the extent to which one evidence statement supports another statement. This view, first presented by Keynes in 1921, has been championed in recent years by Jeffreys [2], among others.

The theory of subjective probability treats probability as a highly personal thing, different for each person. The theory relies on betting situations to obtain probabilities for each person. The idea of using betting in this way was apparently first put forward by Ramsey [6] in 1931.

7.7 I can summarize my attitude with the following comments.

1. There are two main interpretations of probability.
2. Many arguments would be resolved if different words were used for these two interpretations.
3. It is unlikely that different names for probability will gain general acceptance.
4. Both interpretations are useful.
5. The relative frequency concept has been of more use in statistics than the degree of belief interpretation.
6. Recent work in Bayesian statistics has made important advances for the degree of belief interpretation in statistics.

SUMMARY. **Probability** is an undefined term; put another way, we say that it has many different meanings. The two most usual interpretations of the term are (1) an **intrinsic property** of a physical system, and (2) a **measure of belief** in the truth of some statement.

The major attempt to define probability as an intrinsic property regards it as a **limiting relative frequency** in an infinite sequence. The difficulty with this attempt is with the mathematical limit, which lacks some of the properties we would desire. Von Mises attempted to avoid this problem by restricting the class of infinite sequences to those having desirable properties; he called these sequences **collectives**.

Probability as a measure of belief has been developed either in terms of **logical relationships** between statements or in terms of the beliefs unique to individuals, called **subjective** probabilities.

Although no satisfactory definition of probability as a limiting frequency has been given, the phenomenon can be observed by making many trials, such as **tossing a coin**, and observing that the relative frequency tends to settle down around a limiting value.

The single word probability continues to be used to describe rather different ideas, resulting in some controversy. Fortunately, the rules are the same for both interpretations.

REFERENCES

1. Hacking, I. (1965). *Logic of Statistical Inference*, Cambridge, England: Cambridge University Press.
2. Jeffreys, H. (1961). *Theory of Probability*, 3rd Edition, Oxford, England: University Press.
3. Kerrick, J. E. (1946). *An Experimental Introduction to the Theory of Probability*, Copenhagen: J. Jorgensen.
4. Keynes, J. M. (1921). *A Treatise on Probability*, London: Macmillan.
5. Noether, G. E. (1971). *Introduction to Statistics: A Fresh Approach*, Boston: Houghton Mifflin.
6. Ramsey, F. P. (1931). "Truth and Probability," in *The Foundations of Mathematics and Other Logical Essays*, edited by R. B. Brathwaite, London and New York: Routledge & Kegan Paul.
7. Savage, I. R. (1968). Statistics: Uncertainty and Behavior, Boston: Houghton Mifflin.
8. von Mises, R. (1919). "Grundlagen der Wahrscheinlichkeitsrechnung," *Mathematische Zeitschrift*, 5, 52–99.

EXERCISES

1. To become aware of the problem of interpretation, ask several people "What is probability?"
2. I asked a class of 73 students the following question. "I have tossed a regular, U.S. coin 25 times and have obtained 23 heads and 2 tails. What is the probability of heads on the next toss? You may vote for (*a*) greater than one-half, (*b*) less than one-half, (*c*) one-half." The responses were:

Greater than one-half	12
Less than one-half	10
One-half	44
Abstentions	7

   What interpretation do you think was being followed by those who voted for (*a*)? For (*b*)? For (*c*)?
3. What interpretation of probability is being used in the reporting of weather news?
4. In the coin-tossing experiment of Table 7.2 can we say with certainty that the relative frequency of heads will approach one-half as a limit

as n becomes infinite? Can we say that the probability is high that such a thing will happen?

5. What interpretation of probability is being used when odds are stated on the outcome of an athletic event?
6. Toss a die 50 times, recording each time whether a 1 occurs or not. Construct a table similar to Table 7.2 for $n(1)/n$. Then graph the results as in Figure 7.1. Has the ratio settled down as much as the coin-toss ratio after 50 trials?
7. What interpretation (or interpretations) of probability is being used in each of the statements?
   a. The probability of a meltdown at a specified nuclear power plant is .00001.
   b. The probability is .9 that the Notre Dame football team will be ranked among the top 10 teams.
   c. The probability is .05 that there is life on another planet.
   d. The probability of an engine mount failure on an airplane is .001.
   e. The probability of an adequate petroleum supply 20 years hence is .001.
   f. The probability of buying a new car with a major defect is .007.
8. Toss a pair of dice 100 times and record the number of double sixes. Record $n(\text{double sixes})/n$ and graph as in Table 7.2 and Figure 7.1. How many tosses would seem to be necessary in order for the results to settle down in the vicinity of $\frac{1}{36}$
9. An instructor has become embroiled in a dispute over alleged student dishonesty. It has finally been established that the probability that the student actually took all four exams (instead of a standin on some exams) is .9. Then it is discovered that he did not take the first exam. Should the probability that he took the last three be greater or less than .9
10. A friend is willing to bet \$10 against \$5 that team A will win a football game. What is his subjective probability that team A will win?

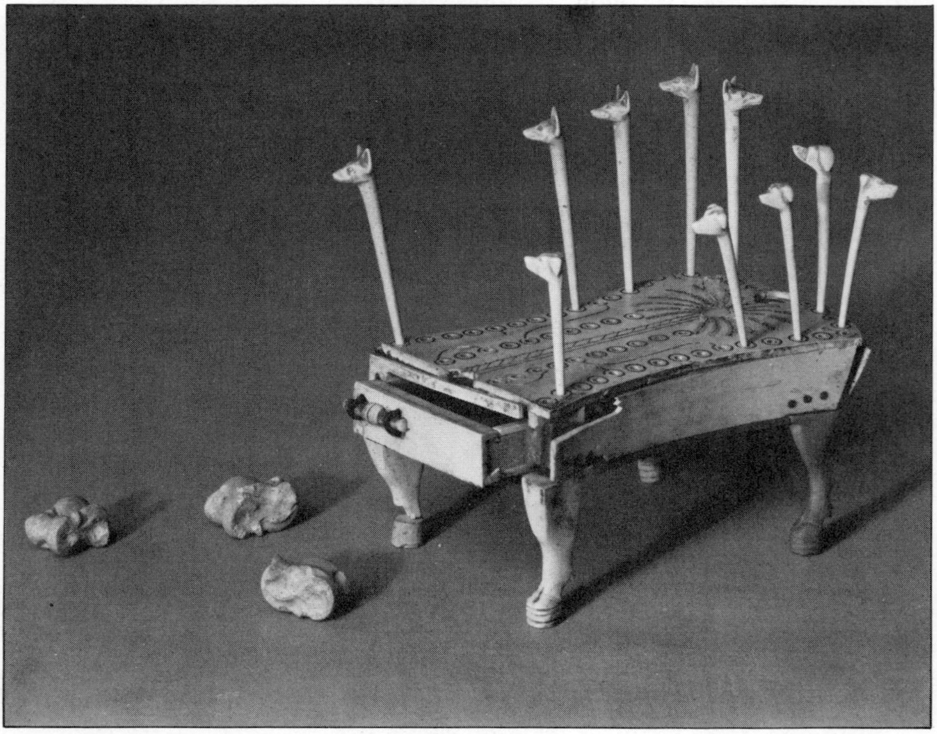

HOUNDS AND JACKALS. Ancient Egyptian game of chance. The Metropolitan Museum of Art, The Carnarvon Collection, Gift of Edward S. Harkness, 1926.

8

RULES OF PROBABILITY

8.1 In the previous lecture we discussed the interpretation of probability, a difficult, philosophical, and controversial matter. Next we take up the rules of probability (the calculus of probability). The rules are easily stated, nonphilosophical, and almost universally accepted. The manner of presentation in vogue in recent years is to state a few **axioms** and then a number of **rules** that result from these axioms. This is often called the **axiomatic theory of probability**. From our point of view, the axiomatic approach completely avoids the important topic of interpretation and concentrates on laying the groundwork for computing probabilities. The axiomatic approach is possible because there is almost complete agreement on the allowable operations with probability.

We will state some definitions and axioms motivated primarily by the relative frequency interpretation, but we will also try to relate these axioms to a degree of belief interpretation.

8.2 Modern set theory provides a convenient vehicle for carrying the necessary definitions. The set of all possible outcomes, called the **sample space**, will be denoted by S. Any *subset* of S will be called an **event**, usually denoted by A, B, C, etc. To illustrate the idea suppose we draw a card from the set of cards consisting of the ace, king, and queen of hearts. There are three possible outcomes, and we denote the sample space by

$$S = \{A, K, Q\}$$

There are eight possible events (subsets) consisting of S, the empty set $\varnothing$, $\{A, K\}$, $\{A, Q\}$, $\{K, Q\}$, $\{A\}$, $\{K\}$, and $\{Q\}$. It may seem strange to the student that $\{A, K\}$ would be listed as an event when we are drawing only one card. However, we would want to be able to discuss the probability that the card is an ace or a king; therefore we need $\{A, K\}$ as an event. Similarly, we need S as an event so that we can discuss the probability of an ace or king or queen. The inclusion of the empty set $\varnothing$ as an event does not follow from intuitive considerations. It is described as an event purely for mathematical convenience. Without its use, some of our manipulations would be cumbersome.

8.3 We now state certain axioms concerning probabilities for any of the events associated with a sample space. These axioms are motivated by the concept of long-run relative frequency, which lies between zero and one. This leads to the first axiom.

Axiom 1. $0 \leq P(A) \leq 1$ for every event A

Since every outcome belongs to the sample space, the long-run relative frequency of S is one. In the coin-tossing experiment of the previous lecture, for example, $n(H \text{ or } T)/n = 1$ for every n. This leads to the second axiom.

Axiom 2. $P(S) = 1$

The next axiom, although not so trivial, follows naturally from relative frequency ideas. Often we ask for the probability that A or B will occur. In the example of the previous section we may ask for P (ace or king). In terms of relative frequency, it seems apparent that

$$\frac{n(\text{aces or kings})}{n(\text{cards})} = \frac{n(\text{aces})}{n(\text{cards})} + \frac{n(\text{kings})}{n(\text{cards})}$$

This leads to the next axiom.

Axiom 3. $P(A \text{ or } B) = P(A) + P(B)$
if A and B are **disjoint** (no points in common)

In terms of **degree of belief**, the idea is that information should be **additive** if A and B have nothing in common. The set theoretic way of saying that A and B are disjoint is to say that $A \cap B = \varnothing$ (the empty set). Also, A or B may be denoted by $A \cup B$. The third axiom is almost always referred to as the **addition rule**.

8.4 Using set theory operations, we can derive several important rules from the three axioms just stated. We will omit the derivations because the rules can be obtained just as directly by thinking about relative frequency ideas. However, it may help the student to think of these rules as resulting from the axioms, just as theorems of geometry result from the given axioms.

Rule 1. $P(\varnothing) = 0$
Rule 2. $P(\bar{A}) = 1 - P(A)$
where $\bar{A}$ is not A

To illustrate Rule 2, the probability of getting an ace on one draw is $\frac{4}{52} = \frac{1}{13}$, while the probability of not getting an ace is $1 - \frac{1}{13} = \frac{12}{13}$.

8.5 Very quickly in the study of probability it seems natural to ask about the occurrence of two or more attributes, such as the card being an ace **and** a spade or the card being an ace **and** red, etc. The probability attached to such a **joint event** is called a **joint probability** and is denoted by $P(AB)$, or $P(A \cap B)$ in set theory symbols. To illustrate the idea, consider the sample space portrayed in Table 8.1.

By inspection, we would conclude

$$P(A) = \frac{4}{52} = \frac{1}{13}$$

$$P(\bar{A}) = \frac{48}{52} = \frac{12}{13}$$

$$P(B) = \frac{13}{52} = \frac{1}{4}$$

$$P(\bar{B}) = \frac{39}{52} = \frac{3}{4}$$

$$P(AB) = \frac{1}{52}$$

$$P(\bar{A}B) = \frac{12}{52}$$

$$P(A\bar{B}) = \frac{3}{52}$$

$$P(\bar{A}\bar{B}) = \frac{36}{52}$$

Table 8.1 Sample Space for Card Problem

	Ace (A)	Not Ace ($\bar{A}$)	Total
Spade (B)	1	12	13
Not spade ($\bar{B}$)	3	36	39
Total	4	48	52

The probabilities involving two events are called **joint probabilities**. The probabilities involving only one event are called **marginal** probabilities, apparently because the marginal probabilities can be obtained from the numbers in the margins of tables such as Table 8.1.

Frequently the sample space is restricted to a subspace. Then the probability is **conditional** on this smaller subspace. When we apply for a life insurance policy, the insurance company may condition the probability of survival on a smaller subspace such as the set of individuals with no record of heart disease. In the card-drawing experiment suppose we ask for the probability of an ace given a spade. We act as though the sample space consisted only of spades, and the desired probability is $\frac{1}{13}$. This is denoted by $P(A \mid B)$, the vertical line being read as "given that."

8.6 Denote the number of points in A by $n(A)$, in B by $n(B)$, and in AB by $n(AB)$. From relative frequency ideas it follows very naturally that

$$\begin{aligned} P(A \mid B) &= \frac{n(AB)}{n(B)} \\ &= \frac{n(AB)}{n(S)} \cdot \frac{n(S)}{n(B)} \\ &= P(AB)/P(B) \end{aligned}$$

The only proviso is that $P(B) \neq 0$. In this manner we also obtain

$$P(B \mid A) = P(AB)/P(A)$$

provided that $P(A) \neq 0$.

8.7 Provided that the **conditional probabilities** exist $[P(A) \neq 0$ and $P(B) \neq 0]$, we can obtain another rule by cross-multiplication.

Rule 3. $P(AB) = P(A \mid B)P(B)$

$$= P(B \mid A)P(A)$$

8.8 The rules of the preceding sections are illustrated by the mortality data of Table 8.2. If we let A be the event that a person lives at least to age 50, then

$$P(A) = .90718$$

Table 8.2 Survivors and Mortality Rate; United States, 1976

Age	Survivors per 100,000 Live Births	Mortality Rate per 1000 Survivors
50	90,718	6.43
51	90,135	7.00
52	89,504	7.62
53	88,822	8.30
54	88,085	9.03
55	87,290	9.77
56	86,437	10.60
57	85,521	11.56
58	84,532	12.71
59	83,458	13.98
60	82,291	15.42

Source. Reproduced from the *Statistical Bulletin, 59,* pp. 8 and 9. Copyright © 1978 by Metropolitan Life Assurance Co. Also reproduced from the National Center for Health Statistics. Reprinted by permission of Metropolitan Life Assurance Co.

The probability of dying before age 50 is given by

$$P(\bar{A}) = 1 - P(A) = .09282$$

Let B be the event that a person dies between age 50 and 51. Then

$$P(B) = (90{,}718 - 90{,}135)/100{,}000$$
$$= .00583$$

In this case $A \cap B = B,$

so

$$P(AB) = P(B) = .00583$$

and

$$P(A \mid B) = P(AB)/P(A)$$
$$= .00643$$

The **mortality rate** at each age, then, is the **conditional probability** of death in the next year given that the person has reached the given age.

8.9 The observant student may have noticed with the card-drawing example in Table 8.1 that $P(A) = P(A \mid B) = \frac{1}{13}$. This simply means that the probability of an ace is not affected by knowing that we have a spade. In an intuitive sense, the two events are **independent**. Note that when $P(A \mid B) = P(A)$ or $P(B \mid A) = P(B)$, Rule 3 gives us $P(AB) = P(A)P(B)$. Accordingly, we give a formal definition of **independence**.

Definition. A and B are independent if and only if $P(AB) = P(A)P(B)$.

The multiplication of probabilities is often called the **product rule**.

8.10 We still need a rule for $P(A \text{ or } B)$. This rule is easily developed by looking at Figure 8.1. Relative frequency ideas suggest that

$$\frac{n(A \cup B)}{n(S)} = \frac{n(A) + n(B) - n(A \cap B)}{N(S)}$$

This leads to the following rule:

Rule 4. $P(A \text{ or } B) = P(A) + P(B) - P(AB)$

As a special case of Rule 4, when A and B are disjoint ($AB = \varnothing$), then $P(AB) = 0$ and $P(A \text{ or } B) = P(A) + P(B)$. In probability literature disjoint events are said to be **mutually exclusive**. The addition of probabilities of mutually exclusive events is usually called the **addition rule**.

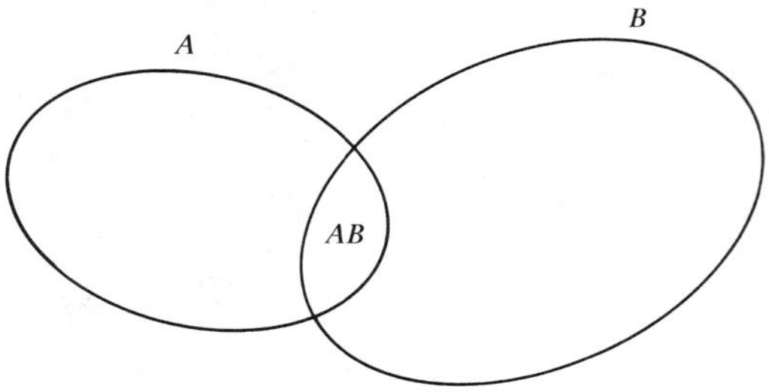

Figure 8.1 Venn Diagram for A or B.

8.11 One particular way of writing a conditional probability has drawn extensive comment. Known as **Bayes' rule** or Bayes' theorem, it simply expresses one conditional probability in terms of other conditional probabilities and marginal probabilities.

Bayes' Rule (Rule 5)

$$P(B \mid A) = \frac{P(A \mid B)P(B)}{P(A \mid B)P(B) + P(A \mid \bar{B})P(\bar{B})}$$

Example. The authorship of books is often a controversial issue among scholars of classical works, and one statistical investigation of a work of disputed authorship was referred to in Lecture 5. Suppose that we have a collection of works, some of which are authored by Paul, and that we have a test for authorship. Let A = indication that Paul authored a given work and B = event that the work was actually authored by Paul. Suppose we have great confidence in our test so that

$$P(A \mid B) = .95 \qquad \text{and} \qquad P(A \mid \bar{B}) = .10$$

If the test indicates for a given work that Paul is the author, we would be interested in the probability that Paul actually is the author given that the test indicates so. That is, we are interested in $P(B \mid A)$. Using Bayes' rule, we obtain

$$P(B \mid A) = \frac{(.95)P(B)}{(.95)P(B) + (.10)P(\bar{B})}$$

In order to go further, we must have $P(B)$. If $P(B)$ is small, say .10,

$$P(B \mid A) = \frac{.95}{.95 + .90}$$

$$= .5135$$

On the other hand, if $P(B)$ is large, say .90,

$$P(B \mid A) = .9844$$

In both cases, an indication by the test that Paul authored the article gives a conditional probability larger than the marginal probability. However, $P(B \mid A)$ is not as large as might have been anticipated if $P(B)$ is small.

8.12 In this chapter we have presented the basic rules for probability calculations. Some of the rules are intuitive; others are not. In the application of even the more intuitive rules there come times when the answers are startling. Probability results obtained from meticulous application of stern rules often do not agree with our intuition, forcing us to sharpen our definitions and to examine our logic.

In the following lecture we will give attention to the problem of probability calculations.

SUMMARY. The **axiomatic** theory of probability takes probability as an undefined quantity satisfying certain **axioms**; then **rules** and theorems are obtained from the axioms.

The set of all possible outcomes, denoted by S, is called the sample space, and any subset A of S is called an event. In the case of two events A and B, the probability of either is called a marginal probability, the probability of both, denoted by $P(AB)$, is called a **joint** probability, and the probability of A with the sample space restricted to the points in B is called the **conditional** probability of A given B, $P(A \mid B)$.

The joint probability can be obtained from the product of a conditional probability and a marginal probability. That is

$$P(AB) = P(B \mid A)P(A)$$

When the joint probability equals the product of marginal probabilities, the events are called independent.

The probability of either A or B is given by

$$P(A \text{ or } B) = P(A) + P(B) - P(AB)$$

When A and B are disjoint, they are called **mutually exclusive**, and the probability of A or B is simply the sum of marginal probabilities.

Bayes' rule allows the calculation of $P(B \mid A)$ given $P(A \mid B)$, $P(A \mid \bar{B})$, $P(B)$, and $P(\bar{B})$. That is, it gives one conditional probability in terms of another conditional probability and the appropriate marginal probabilities.

REFERENCES

1. Bulmer, M. G. (1965). *Principles of Statistics*, Cambridge, Mass.: The M.I.T. Press.
2. Cramér, Harold. (1955). *The Elements of Probability Theory and Some of Its Applications*, Princeton, N.J.: Princeton University Press.
3. King, Amy C., and Cecil B. Read. (1963). *Pathways to Probability*, New York: Holt, Rinehart and Winston.

4. Jeffreys, H. (1961). *Theory of Probability*, 3rd Edition, Oxford, England: Clarendon Press.
5. Metropolitan Life Assurance Co. (1978). *Statistical Bulletin, 59.*
6. Mosteller, R., R. E. K. Rourke, and G. B. Thomas, Jr. (1961). *Probability and Statistics*, Reading, Mass.: Holden-Day.
7. Parzen, E. (1960). *Modern Probability Theory and Its Applications*, New York: Wiley.

EXERCISES

1. A ball is drawn from an urn containing four balls numbered 1, 2, 3, 4. Give the sample space and all of the 16 possible events.
2. Two balls are drawn from the urn of Exercise 1, one at a time, without replacement of the first ball before the second ball is drawn. List the six possible outcomes in the sample space and the 64 possible events.
3. For the sample space in Table 8.1 calculate

$$P(A \mid B), P(A \mid \bar{B}), P(B \mid A), \text{ and } P(B \mid \bar{A}).$$

4. From the data given in Table 8.2 verify the mortality rates for ages 50 to 59.
5. From the data of Table 8.2 calculate the probability of surviving until age 61.
6. What interpretation of probability is being used in probability statements based on Table 8.2?
7. An urn contains three balls numbered I, II, and III. Two balls are drawn at random from the urn, one at a time, without replacement of the first ball, before the second ball is drawn.
   a. List the six points in the sample space, each point specifying an outcome for the first and second ball drawn.
   b. Let A be the event that ball I is drawn on the first draw and determine $n(A)$. What is $P(A)$?
   c. Let B be the event that ball I is drawn on the second draw and determine $n(B)$. What is $P(B)$?
   d. Let C be the event that ball II is drawn on the first draw and determine $n(C)$. What is $P(C)$?
   e. What is $P(AC)$?
   f. What is $P(A \mid C)$?

8. Work Exercise 7 supposing that the first ball drawn is replaced before the second ball is drawn. For part a there will be nine points in the sample space.

9. Suppose that in Exercise 7, balls I and II are white and III is red. Let A = event that first ball drawn is white and B = event that second ball drawn is white. Determine:
   a. $P(A)$.
   b. $P(B)$.
   c. $P(AB)$.
   d. $P(A \mid B)$.
   e. $P(B \mid A)$.

10. Two cards are drawn (without replacement) at random from the four aces of a bridge deck.
    a. List the 12 points in the sample space.
    b. Let A be the event of getting the ace of spades. List the points in A and determine $P(A)$.
    c. Let B be the event of getting the ace of hearts. List the points in B and determine $P(B)$.
    d. Determine $P(A \mid B)$, the probability of the ace of spades given the ace of hearts.
    e. Determine the probability of the ace of spades given that we obtain at least one red ace.

11. Show that $\varnothing$ is independent of any event A. Also show that $\varnothing$ and A are mutually exclusive.

12. A fair coin is tossed twice. Let A be the event of heads on the first toss and B be the event of heads on the second toss. Show that A and B are independent but not mutually exclusive.

13. If we are told that two events are independent, can we conclude that they are not mutually exclusive?

14.
$$P(A \text{ or } B) = \tfrac{7}{16}$$
$$P(A) = \tfrac{1}{4}$$
$$P(B) = \tfrac{1}{4}$$

    Are A and B mutually exclusive?

15.
$$P(A \mid B) = \tfrac{1}{6} \qquad P(\bar{B}) = \tfrac{1}{2} \qquad P(B \mid A) = \tfrac{1}{4}$$

    Find $P(A)$ and $P(B)$.

16. We are given that

$$P(A) = \tfrac{7}{16} \qquad P(B) = \tfrac{8}{16}$$
$$P(C) = \tfrac{6}{16} \qquad P(AB) = \tfrac{3}{16}$$
$$P(AC) = \tfrac{3}{16} \qquad P(BC) = \tfrac{3}{16}$$
$$P(ABC) = \tfrac{2}{16}$$

a. $P(A \text{ or } B \text{ or } C) = ?$
b. Sketch the Venn diagram showing probabilities for this problem.

17. We are given that

$$P(A) = P(B) = P(C) = \tfrac{1}{2}$$
$$P(AB) = P(AC) = P(BC) = \tfrac{1}{4}$$
$$P(ABC) = 0$$

a. Sketch the Venn diagram and show the probabilities for this problem.
b. Show that A, B, and C are pairwise independent but not jointly independent.

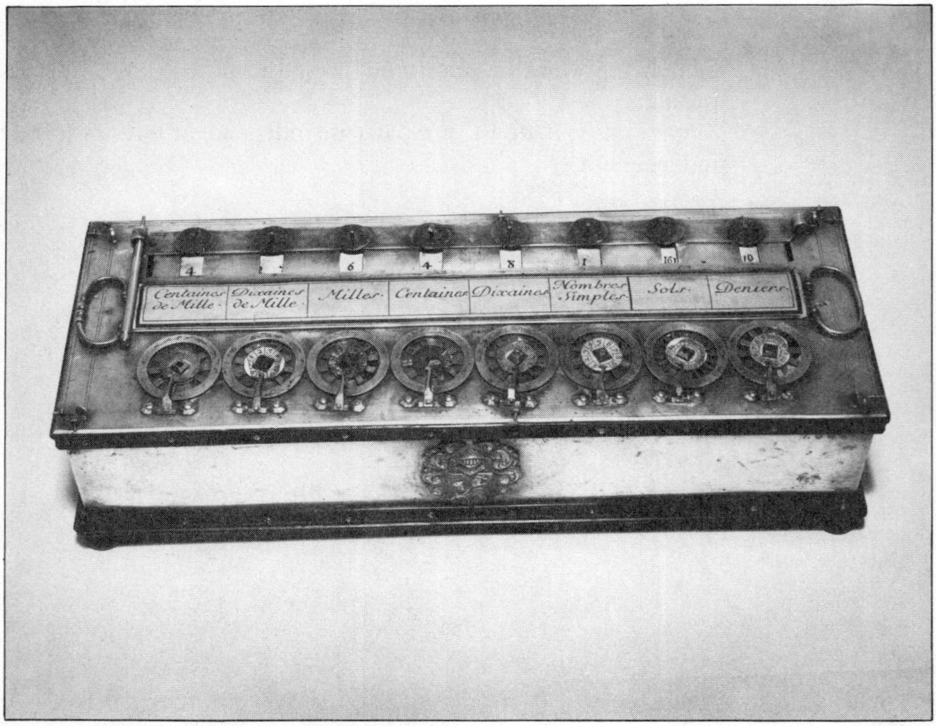

PASCAL'S CALCULATING MACHINE. Courtesy of IBM.

9

COMPUTATION OF PROBABILITY

9.1 The so-called **classical definition** of probability states that if an experiment can result in n **equally likely** outcomes and if $n(A)$ of these outcomes possess the property A, the probability of A is given by

$$P(A) = \frac{n(A)}{n}$$

Of course, this is not really a definition because it contains the term "equally likely," which presumes a meaning of probability. Nevertheless, it does provide a useful way of formalizing our thinking about probability. In this lecture we will present a number of examples where all points in the sample space are taken to be equally likely.

Example 1. An experiment consists of drawing two balls in succession, without replacement of the first ball, from an urn containing two white balls and one red ball. We are asked for the probability that the second ball is white. For this example, let us number the balls with the corresponding colors as follows.

Ball	Color
1	W
2	W
3	R

The sample space in terms of ball number is as follows.

Outcome	First Ball	Second Ball
1	1	2
2	1	3
3	2	3
4	2	1
5	3	1
6	3	2

The event A, consisting of the outcomes where the second ball is white, is the following.

Outcome	First Ball	Second Ball
1	1	2
4	2	1
5	3	1
6	3	2

Thus the probability that the second ball is white is $n(A)/n = \frac{4}{6} = \frac{2}{3}$. This answer may be counter to the intuition of students, who unconsciously think in terms of conditional probability.

In this example it is a simple matter to list all of the points in the sample space and to count the points belonging to the event A. For more complex problems the counting may become an onerous task unless we make use of counting formulas and counting algorithms.

9.2 Mathematics abounds with fascinating counting problems, and the solution to some of these problems requires the highest ingenuity. However, many problems can be solved with elementary combination and permutation formulas. The arrangements of balls of Example 1 are called **permutations** because the order is important. That is, the arrangement 1, 2, 3 is considered to be different than 3, 2, 1. By contrast, the arrangement of cards in a bridge hand is a **combination** because the order of cards is not important—the value of the hand not depending on the sorting. It was mentioned in Lecture 2 that the word permutation was first mentioned by Bernoulli. The word combination was first used in the modern sense in 1666 by Leibniz in his "Dissertation Concerning the Combinatorial Arts."

The number of combinations of n things taken r at a time is denoted by $\binom{n}{r}$ and is given by

$$\binom{n}{r} = \frac{n!}{r!(n-r)!}$$

where $n!$ (n factorial) is given by $n! = n(n-1)(n-2)\ldots 1$ with the usual convention that $0! = 1$.

The number of permutations of n things taken r at a time is given by $P(n, r) = n!/(n-r)!$. To illustrate the use of this formula, consider Example 1. The number of points in S is obtained from $P(3, 2) = 3!/1! = 6$.

Example 2. An experiment consists of dealing 13 cards from a **well-shuffled** bridge deck. What is the probability that the hand contains the ace of spades and the ace of hearts? To solve this problem, note that we need to calculate the ratio $n(A)/n$, and the major problem is a counting problem. Now n is simply the number of ways of choosing 13 cards from the deck, $\binom{52}{13}$. The $n(A)$ numerator requires a little more thought. In every hand containing the two specified aces, there are 11 other cards that can be selected in $\binom{50}{11}$ ways. Therefore the probability is

$$\binom{50}{11} \div \binom{52}{13} = \left(\frac{50!}{11!\,39!}\right)\left(\frac{13!\,39!}{52!}\right)$$

$$= \frac{(13)(12)}{(52)(51)} = \frac{1}{17}$$

9.3 To summarize these ideas, many of the simple problems can be conceptualized by thinking of the list S of all the possible outcomes judged to be equally likely and of the subset A in this list. The desired probability is then obtained by forming the ratio of the number of points in A, $n(A)$, to the number of points in S, n. At this point the need arises for counting devices; this need is partially met by combinations and permutations.

Confusion usually arises for beginning students (and for me, for that matter!) about whether to use combination or permutation formulas.

This confusion can be avoided if we think carefully about the points in S, the sample space. If they are permutations, naturally we use permutation formulas; if they are combinations, we use combination formulas.

9.4 The famous **birthday problem** is concerned with the probability that when r people are selected at random from a large population, at least two have the same birthday (see Feller [2] and Munford [3] for a recent article). Ignoring leap years, the sample space is the set of all possible r-tuples $(x_1, x_2, \ldots, x_r)$, where each of the xs can equal $1, 2, 3, \ldots, 365$. Then the number of points in the sample space is

$$n(S) = (365)^r$$

Let A be the event that at least two people have the same birthday; $\bar{A}$ is the event that all birthdays are different. $\bar{A}$ is the set of all permutations of 365 dates taken r at a time. Thus

$$n(\bar{A}) = P(365, r)$$

Then

$$P(\bar{A}) = P(365, r)/(365)^r$$

and

$$P(A) = 1 - P(\bar{A})$$

We can evaluate $P(A)$ for various values of r. For $r = 23$, $P(A)$ exceeds one-half. Thus, for 23 or more people, the odds are in favor of at least two people having the same birthday. It has been shown by several people (e.g., Munford [3]) that if all birthdays are not equally likely, $P(A)$ is increased.

9.5 The Society for Psychical Research was founded in London in 1882 because of the conviction there was a need for scientific study of paranormal events (odd and unexplained kinds of events). The term **parapsychology** is used to describe experimental psychical research. A review of experimental investigations of **extrasensory perception** is given by Thakur [4] and Thouless [5]. Since the early card-guessing experiments, many experiments have been devised to test the hypothesis that extrasensory perception does exist. Regardless of the outcome, there is always the question of whether the results could be explained by chance. In fact, one of the chapters in the book by Thouless is entitled "Can the Apparent Successes be Explained by Chance?"

A psychic deals the face cards from spades and hearts, face up, in a random order, looking at the cards and concentrating on their color. In another room a colleague who knows there are four red and four black cards tries to guess the color of the card drawn on a signal from an impartial observer that a card has been dealt. If six out of eight cards are correctly identified by color, would you feel that this supports the claim of telepathy?

The possible outcomes of the experiment are the possible records of eight guesses with four red and four black. We can find the number of points in the sample space, $n(S)$, by thinking of the number of ways that we can assign red to four of the guesses (black going to the other four). This is simply the number of combinations of eight things taken four at a time. So

$$n(S) = \binom{8}{4} = 70$$

Let us suppose that all 70 outcomes are equally likely and find the probability of identifying correctly six or more cards.

In order to identify correctly six cards by color, three red cards and three black cards must be identified. The number of ways to identify correctly three red cards is $\binom{4}{3}$ and for each of these ways, three black cards can be identified in $\binom{4}{3}$ ways. So the number of ways to identify six cards is the product $\binom{4}{3}\binom{4}{3} = 16$. Since there is only one way to identify correctly all eight cards, the probability of identifying correctly six or more cards is $\frac{17}{70}$.

Even if the colleague is guessing, there is almost a 25% chance of identifying correctly six or more cards in this experiment.

9.6 According to Duncan [1], it was only in the 1920s that statistical theory began to be applied effectively to **quality control**. Walter A. Shewhart of the Bell Telephone Laboratories began the work and in 1931 published a book entitled *Economic Control of Quality of Manufactured Product.* Following Shewhart's visit to London in 1932, the development of statistical quality control in Britain was swift. Although the development lagged in the United States, it received a major impetus during World War II. Duncan reviews the development of **acceptance sampling plans** from the Dodge-Romig Tables up to the present.

The simplest acceptance sampling plan works as follows. From a manufactured lot of N items we take a random sample of n items and accept the lot if the number of defective items is less than or equal to a specified number, c. For example, we might accept a lot of 10 items if we found three defectives or fewer in a sample of size 5.

Anyone can devise an acceptance sampling plan. The next problem is to determine the properties, or **operating characteristics**, of the plan devised. We might try to determine the probability of accepting lots of varying quality. If the lot of 10 items contains four defective items, what is the probability of accepting the lot?

The number of points in the sample space is the number of combinations of 10 things taken five at a time, $\binom{10}{5} = 252$. The number of samples with three defectives or fewer is

$$\binom{6}{2}\binom{4}{3} + \binom{6}{3}\binom{4}{2} + \binom{6}{4}\binom{4}{1} + \binom{6}{5}\binom{4}{0} = 60 + 120 + 60 + 6$$
$$= 246$$

Therefore the probability of accepting the lot containing four defective items is $\frac{246}{252} = .976$.

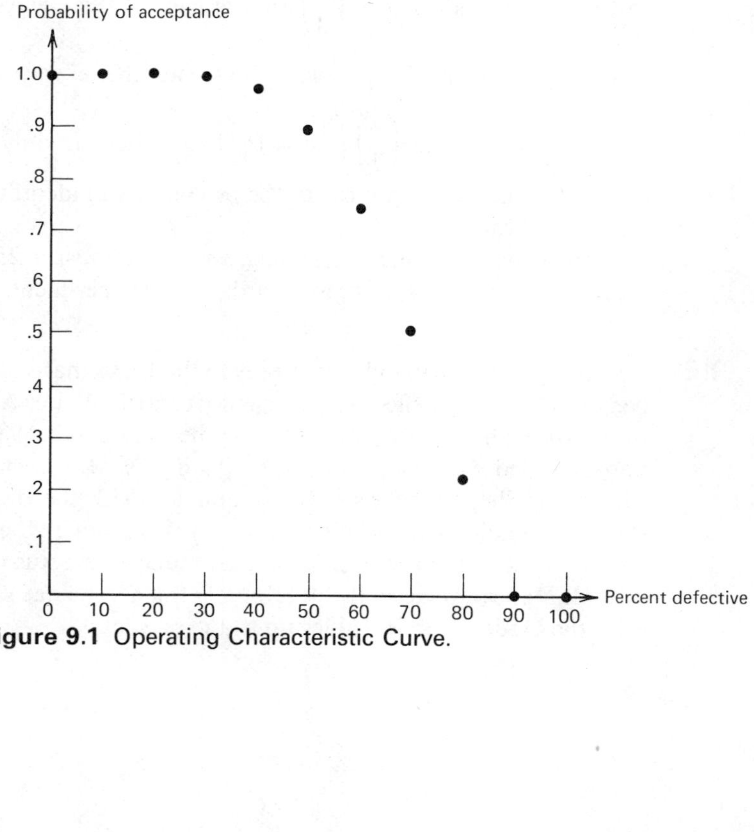

Figure 9.1 Operating Characteristic Curve.

Similarly, we can calculate the probability of acceptance for any number of defectives in the lot and plot the probability of acceptance versus percent defective in the lot. Such a curve is called an **operating characteristic curve** (OC curve), and the OC curve for the present example is given in Figure 9.1.

9.7 With the very high standards required by the development of missile systems and space travel, extensive development of **reliability** engineering has occurred in the last 25 years. Suppose three electrical components A, B, and C are connected in series, as in Figure 9.2.

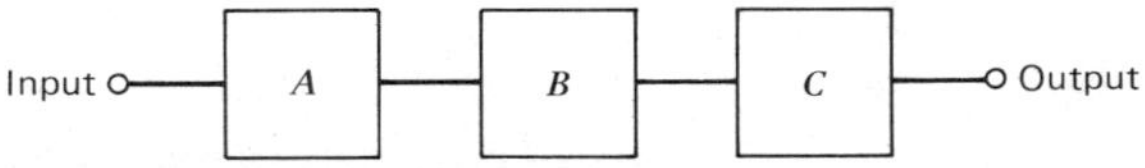

Figure 9.2 Components in Series.

In order for the system to work all three components must function. If the component probabilities are $p_A = .95$, $p_B = .90$, and $p_C = .99$, the probability of the system working (the reliability) is $(.95)(.90)(.99) = .84645$. One way of increasing the system reliability is to build a parallel system, as in Figure 9.3. Then the system reliability is

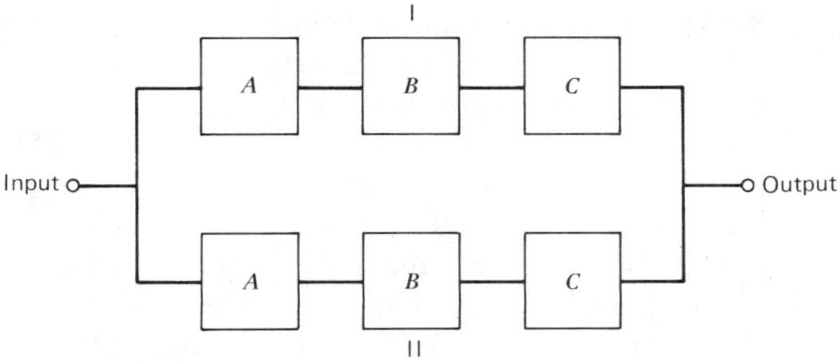

Figure 9.3 Parallel Systems.

$$
\begin{aligned}
P(\text{I or II}) = P(\text{I}) + P(\text{II}) - P(\text{I and II}) \\
&= .84645 + .84645 - (.84645)^2 \\
&= .97642
\end{aligned}
$$

Alternatively, we can use parallel circuits, as in Figure 9.4.

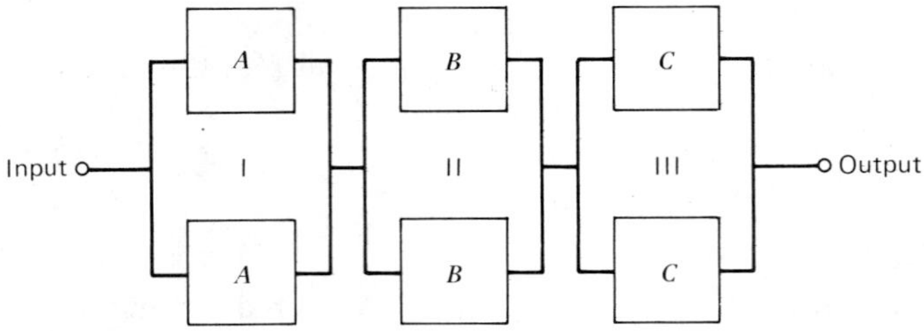

Figure 9.4 Parallel Circuits.

For the system in Figure 9.4, the system reliability is

$$P(\text{I and II and III})$$

But

$$P(\text{I}) = .95 + .95 - (.95)^2 = .9975$$

$$P(\text{II}) = .90 + .90 - (.90)^2 = .99$$

and

$$P(\text{III}) = .99 + .99 - (.99)^2 = .9999$$

Therefore the system reliability is .98743. In this case we see, therefore, that the system in Figure 9.4 is more reliable than the one in Figure 9.3.

9.8 We often have opportunities to make use of simple **tree diagrams** in the solution of probability problems. We will not give a general description of a tree diagram, but we will give an example. Suppose that we are asked for the probability of exactly two aces in the draw of three cards from a bridge deck. We can imagine the cards being drawn sequentially, and the possibilities on each draw are ace or not-ace. We can then visualize the possible outcomes as in the tree in Figure 9.5. The probabilities shown on the branches are conditional on the previous draw (draws) and should be verified. The probability of climbing from the base to any of the branches is obtained by multiplying the probabilities. Thus the probability of climbing to branch 1 is obtained by

$$\frac{1}{13} \times \frac{3}{51} \times \frac{2}{50} = \frac{1}{(13)(17)(25)}$$

The probability of getting exactly one ace in three draws is the sum of probabilities for branches 4, 6, and 7 and is given by $(\frac{1}{3})(\frac{48}{51})(\frac{47}{50}) + (\frac{12}{13})(\frac{4}{51})(\frac{47}{50}) + (\frac{12}{13})(\frac{47}{51})(\frac{4}{50})$.

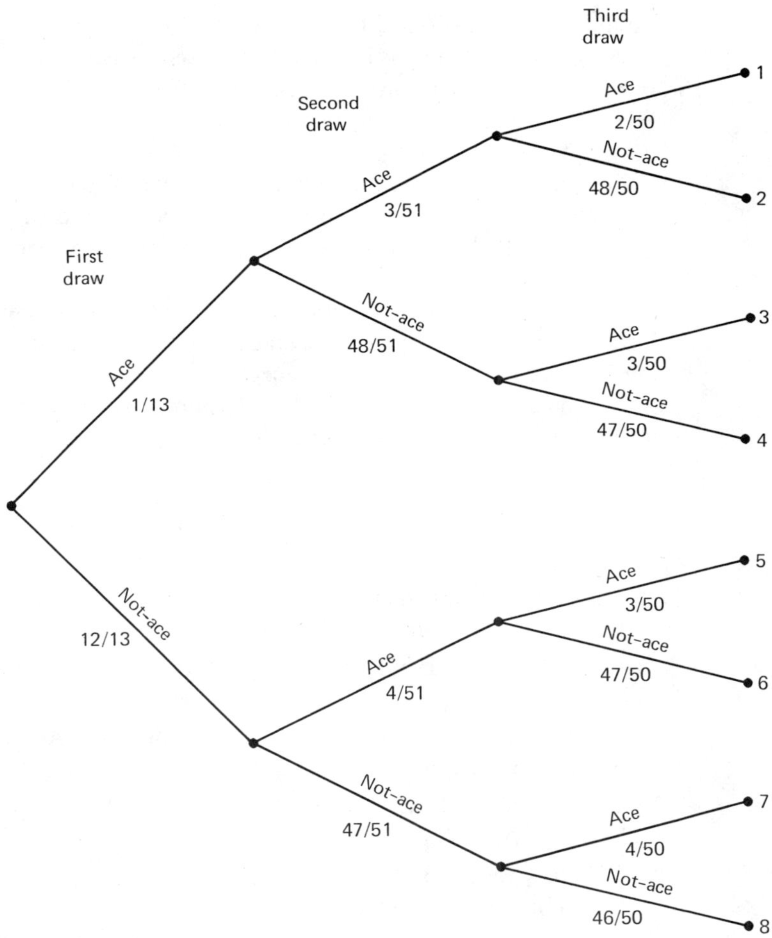

Figure 9.5 Tree Diagram: Card Problem.

9.9 In conclusion, we should mention that obtaining a solution to a simply stated probability problem can be extremely difficult and students may be discouraged by initial difficulties. These difficulties should not be allowed to obscure the conceptually simple process involved.

SUMMARY. In many probability problems it seems reasonable to assume that all points in the sample space are **equally likely**, or equally probable. Then the probability of an event, *A*, is simply the number of points in A, $n(A)$, divided by the number of points in S, n. That is, $P(A) = n(A)/n$. This formula is known as the **classical definition** of probability, but it is not a definition at all. In these problems the calculation of a probability is essentially a **counting problem**. The number of points in A and in S must be counted.

For purposes of counting, the **combination** and **permutation** formulas are invaluable. Once the points in S have been listed or even partially listed, it will be clear whether the points are permutations or combinations.

The calculation of probabilities is illustrated with several examples, including an **urn** problem, the **birthday** problem, **extrasensory perception**, **quality control**, and **reliability theory**.

Tree diagrams are also useful for obtaining probabilities, particularly for events consisting of sequential trials.

REFERENCES

1. Duncan, Acheson J. (1974) *Quality Control and Industrial Statistics*, Homewood, Ill.: Richard D. Irwin.
2. Feller, William. (1950). *Probability Theory and Its Applications*, New York: Wiley.
3. Munford, A. G. (1977). "A Note on the Uniformity Assumption in the Birthday Problem," *The American Statistician*, *31*, 119.
4. Thakur, Shivesh C. (1976). *Philosophy and Psychical Research*, London: Allen & Unwin.
5. Thouless, Robert H. (1963). *Experimental Psychical Research*, Baltimore: Penguin Books.

EXERCISES

1. Evaluate the probability that two out of r people have the same birthday for $r = 5, 10, 15, 20, 23, 25, 30, 35$. Graph the probability versus r. In several of your social groups inquire about people with matching birthdays. Are the results consistent with the theoretical answers?
2. Conduct the color-guessing experiment described in Section 9.5. Does your experimental data support the mental telepathy experiment?

3. Verify the OC curve of Figure 9.1 by calculating the probability of acceptance for $p = .1, .2, \ldots, .9$.
4. Show that $\binom{n}{r} = \binom{n}{n-r}$.
5. Show that $\binom{n-1}{r} + \binom{n-1}{r-1} = \binom{n}{r+1}$.
6. Determine the OC curve for an acceptance sampling plan that accepts the lot with two defective items or fewer in a sample of four items from a lot of size 15.
7. A 13-card bridge hand is drawn at random from a well-shuffled bridge deck.
   a. Let A be the event that the hand contains two aces. What is $P(A)$?
   b. Let B be the event that the hand contains three aces. What is $P(B)$?
   c. What is the probability that the hand contains four aces?
8. To continue with Exercise 7, suppose we are not interested in bridge hands but in the outcome for the first card drawn, the second card drawn, etc.
   a. Are the points in the sample space combinations or permutations?
   b. What is the number of points in the sample space?
   c. Let A be the event that the first card drawn is a spade. Determine $n(A)$ and $P(A)$.
9. If each of the components in Figure 9.2 has a reliability of 0.99, what is the system reliability?
10. With 10 components in series as in Figure 9.2, what must the component reliability be in order for the system reliability to be 0.90?
11. With three components each of A, B, and C with component reliabilities, respectively, of 0.95, 0.90, and 0.99, how can we achieve the greatest system reliability? Should we use three parallel systems as in Figure 9.3 or should we use the components in parallel as in Figure 9.4?
12. Users of books in a library are asked not to reshelve their own books. If a book is borrowed from a shelf holding 50 books and returned to a random place on that same shelf, what is the probability that it will be in the proper location?
13. If two of the books (see Exercise 12) are returned to random places, what is the probability that both books are in the proper location?

14. A two-stage acceptance sampling plan for lot size of 10 proceeds as follows. An initial sample of size 3 is drawn. If there are no defective items the lot is accepted. If there are two or three defective items the lot is rejected. If there is only one defective item, another two items are drawn. If no additional defective items are found the lot is accepted. Otherwise it is rejected. Determine the *OC* curve for $p = .1$, $.2, \ldots, .9$.
15. A coin is tossed five times. Calculate the probability of two or more heads.
16. A woman living outside a large city drives her car to the nearest railroad station, parks it there, and takes the train into town to attend the opera. Once there she becomes concerned that she may have left the headlights on. If so, the car battery will be down by the time she gets back to the station after the opera. She estimates that the probability that the lights are on is .9. She further estimates that the probability of being able to go back to the station, check the lights, and return in time for the opera is .5. What should she do if she wishes to maximize the probability of getting home in her car **and** attending the opera?
17. A homeowner's lawn mower has become worn and sometimes will not start again as long as the motor is hot. To mow his lawn, he must kill the motor and empty the grass catcher four times. The probability that the mower will start again if he kills the motor is .9, and the probability that he can mow the whole lawn without running out of gasoline is .8. If he leaves the mower stopped long enough to refill it, the probability of its starting again is .7. Should he plan on refilling the mower on one of the stops or not?
18. A car owner is debating on whether or not to order spark plug wires from an auto parts catalog. He judges that the probability that new wires will cure his ignition problems is .3. If spark plug wires are not the problem, he plans to take the car to a garage for a tune-up. What should he do to minimize his expected cost if the spark plug wires cost \$18 and the tune-up will cost \$80?
19. It cost \$9 for each spraying of a large, papershell pecan tree. If all three sprayings are at the right time, the crop will be 100 pounds; if there are two sprayings at the right time, it will be 60 pounds; if there is one spraying at the right time, it will be 30 pounds; if the tree is not sprayed, there is no crop. The probability for each spraying being at the right time is .9. What is the expected net return if pecans are worth \$2 per pound?

THREE DISTRIBUTIONS

JAMES BERNOULLI. 1654–1705. Smith Collection, Rare Book and Manuscript Library, Columbia University.

10

THE BINOMIAL DISTRIBUTION

10.1 *And strange to tell, among that Earthen Lot*
Some could articulate, while others not:
And suddenly one more impatient cried—
*"Who is the Potter, pray, and who the Pot?"**

Edward Fitzgerald's translation of the *Rubaiyat* of Omar Khayyam [2] seems appropriate as a starting point for a discussion of the **binomial distribution**. The **binomial expansion** and the **binomial coefficients** have been known to mathematicians for hundreds of years but have been rediscovered again and again until it seems impossible to say who is the "potter" and who is the "pot." Interestingly, Omar Khayyam himself may have been familiar with the binomial expansion and coefficients. Before giving a few traces of the origin of the binomial expansion, we should review a few facts about the expansion itself.

10.2 In elementary algebra courses students learn that

$$(a + b)^2 = a^2 + 2ab + b^2$$

and

$$(a + b)^3 = a^3 + 3a^2b + 3ab^2 + b^3$$

By multiplying $(a + b)$ times the expanded form of $(a + b)^3$, we obtain

$$(a + b)^4 = a^4 + 4a^3b + 6a^2b^2 + 4ab^3 + b^4$$

* Reprinted by permission of the publishers George Allen & Unwin Ltd. from *The Romance of the Rubaiyat* by A. J. Arberry. Copyright © 1959 by George Allen & Unwin Ltd., London.

At this point students may be motivated to memorize some rule to generate the terms of the expansion more easily. It was observed long ago that the sum of the exponents is constant from term to term. That is, if Ca^xb^y is a term in the expansion of $(a + b)^n$, $x + y$ always equals n. What is needed is a simple expression for the coefficient. It was also discovered long ago that the coefficient of the term involving a^x was $\binom{n}{x}$, the number of combinations of n things taken x at a time. Then the binomial expansion can be written as

$$(a + b)^n = \sum_{x=0}^{n} \binom{n}{x} a^x b^{n-x}$$

10.3 **Pascal's triangle** provides a simple device for generating the binomial coefficients; the rows of the triangle correspond to different values of n. The elements in each row are the coefficients of the terms in the expansion for that value of n. The enticing property of the array is that any coefficient (other than 1) is the sum of the two coefficients to the immediate left and right in the row above it. So, for $n = 8$, we would obtain the coefficients $1, 8 = 1 + 7, 28 = 7 + 21, 56 = 21 + 35, 70 = 35 + 35, 56 = 35 + 21, 28 = 21 + 7, 8 = 7 + 1$, and 1.

According to Hogben [4, p. 162], Khayyam may have been familiar with the binomial expansion as early as 1000 A.D. In Lecture 2 we called attention to the reference by Smith, tracing the triangle to a Chinese writer as early as 1303. Eves [1] points out that the binomial coefficients were given by Michael Stifel in *Arithmetica integra*, published in 1544. Pascal's name seems to have been firmly attached to the triangular array of coefficients about 1665, although a triangle-type array was also given by Bernoulli in the 1713 *Ars Conjectandi.*

10.4 The binomial probability distribution is often attributed to Bernoulli (e.g., see Kendall and Buckland [5]). In view of the murky origins of the binomial expansion, such a singular source for its use in probability seems suspect. However, it is clear that Bernoulli did use the binomial expansion for probability problems.

The binomial expansion arises from consideration of the following problem. Several (n) **independent trials** are to be performed with **two** possible outcomes (a **success** or a **failure**) on each trial. If the probability of a success on each trial is p, what is the probability of exactly x successes? For example, in five ($n = 5$) tosses of a die, what is the

Table 10.1 Illustrations of Binomial Probabilities[a]

Event	Trials 1	2	3	4	5	Probability of Event
1	4	4	4	a	a	$(\frac{1}{6})^3(\frac{5}{6})^2$
2	4	4	a	4	a	$(\frac{1}{6})^3(\frac{5}{6})^2$
3	4	4	a	a	4	$(\frac{1}{6})^3(\frac{5}{6})^2$
4	4	a	4	4	a	$(\frac{1}{6})^3(\frac{5}{6})^2$
5	4	a	4	a	4	$(\frac{1}{6})^3(\frac{5}{6})^2$
6	4	a	a	4	4	$(\frac{1}{6})^3(\frac{5}{6})^2$
7	a	4	4	4	a	$(\frac{1}{6})^3(\frac{5}{6})^2$
8	a	4	4	a	4	$(\frac{1}{6})^3(\frac{5}{6})^2$
9	a	4	a	4	4	$(\frac{1}{6})^3(\frac{5}{6})^2$
10	a	a	4	4	4	$(\frac{1}{6})^3(\frac{5}{6})^2$

[a] "Not four."

probability of exactly three ($x = 3$) fours? The 10 mutually exclusive events with their associated probabilities are given in Table 10.1.

Now the probability of exactly three fours is the sum of the probabilities of the mutually exclusive events producing exactly three fours. So

$$P(x = 3) = 10(\tfrac{1}{6})^3(\tfrac{5}{6})^2$$

Every line in Table 10.1 has the same probability. If we can ascertain the number of lines we do not need to list the events. It is easily seen that the number of lines is simply the number of ways that we can pick three positions from the possible five, $\binom{5}{3}$. The desired probability is then

$$P(x = 3) = \binom{5}{3}(\tfrac{1}{6})^3(\tfrac{5}{6})^{5-3}$$

Reasoning in this way we can conclude that the probability of exactly x successes in n trials is generally

$$P(x \text{ successes}) = \binom{n}{x}p^x(1-p)^{n-x}$$

$$= \binom{n}{x}p^x q^{n-x}$$

where

$$q = 1 - p$$

It is quickly noted that this expression resembles one of the terms of a binomial expansion. It is, in fact, one of the terms of the binomial expansion of $(p + q)^n$. For this reason it is known as the **binomial probability distribution** or, more simply, the **binomial distribution**.

It should also be pointed out that we could have arrived at the binomial probability formula by use of a tree diagram, as in Lecture 9. The tree would have 2^n branches, and the probability for x successes is the sum of the probabilities associated with $\binom{n}{x}$ branches.

10.5 Tables of the binomial distribution have existed for many years. Table A.2 gives values up to $n = 10$ for several values of p. For example, we can read directly from the table that the probability of three successes from eight trials with $p = \frac{1}{3}$ is .2731. If we throw a die eight times, the probability of getting a one or a two on three throws of the die is .2731.

10.6 Gridgeman [3] uses the binomial distribution to analyze what is surprising about a family of 11 children—five girls followed by six boys. He refers to this example, which is given in the Life Science Library, as a very rare grouping with a probability of $\frac{1}{2048}$. As Gridgeman points out, any family of 11 children (with boys and girls in any specified order) has exactly the same probability—$(\frac{1}{2})^{11} = \frac{1}{2048}$.

Given that there are 11 children, it is not surprising that there are five boys and six girls, because the five-six split is more likely than any other. However, given five boys and six girls, the probability of five boys followed by six girls or six girls followed by five boys is $\frac{1}{231} = 2/\binom{11}{5}$, an unlikely event but with considerably higher odds than $\frac{1}{2048}$.

10.7 In Lecture 8 an experiment in color guessing was described for which the probability of identifying correctly six or more cards was $\frac{17}{70}$. Suppose the experiment were repeated 10 times, with six or more cards identified correctly 8 out of the 10 times. Is this surprising? With $n = 10$ and $p = \frac{17}{70}$, we want to find $P(x = 8) + P(x = 9) + P(x = 10)$. Since $\frac{17}{70}$ is approximately .25, let us use Table A.2 to find this probability. We see that it is approximately .0004 + .0000 + .0000 = .0004. Certainly this sort of success would be so unlikely to have occurred by chance that it would give strong support for mental telepathy.

10.8 The **mean** and **variance** are easily found by algebraic methods to be

$$\mu = np$$

and

$$\sigma^2 = npq$$

These **moments** also agree with our intuitive interpretations, as described in Lecture 6. The mean is the average or expected number of successes, which is just the number of trials multiplied by the probability of success. This calculation is almost a subconscious one in many cases. If we toss a coin nine times we would say the expected number of heads is $\frac{9}{2}[np = (9)(\frac{1}{2})]$. The variance is not quite as intuitive. However, we can see from Table A.2 that the variance increases as n increases and as p approaches one-half. A little thought will convince us that p and q should enter equally into the variance formula so that the expression npq is reasonable.

To indicate how the algebraic derivation goes, let $n = 2$. The frequency distribution is as follows.

x	f
0	$(1-p)^2$
1	$2p(1-p)$
2	p^2

Then

$$\begin{aligned}\mu &= \Sigma f_i x_i \\ &= 2p(1-p) + 2p^2 \\ &= 2p\end{aligned}$$

and

$$\begin{aligned}\sigma^2 &= (0-2p)^2q^2 + (1-2p)^2 2pq + (2-2p)^2p^2 \\ &= 4p^2q^2 + (1-4pq)2pq + 4p^2q^2 \\ &= 2pq(4pq + 1 - 4pq) \\ &= 2pq\end{aligned}$$

10.9 We have shown how the binomial distribution arises from the concepts of the previous lectures, given a few of its properties, and illustrated its use in calculations. Do not be misled into thinking that it is useful only for card-drawing experiments. Knowledge of the binomial distribution has applications in genetics, acceptance sampling, reliability, and many other areas.

SUMMARY. The **binomial distribution** derives its name from the familiar **binomial** expansion of $(a + b)^n$. The **binomial coefficients** of this expansion have the form $\binom{n}{x}$.

When n independent trials are performed with **success** or **failure** possible on each trial and a **constant probability** of **success** p, the **probability of x successes** is given by

$$P(x \text{ successes}) = \binom{n}{x} p^x (1 - p)^{n-x}$$

This is one of the terms in the binomial expansion of $(q + p)^n$.

The **mean** and **variance** of the binomial distribution are np and $np(1 - p)$, respectively.

For large n the calculation of binomial probabilities becomes tedious, but **tables** of the distribution and pocket calculators have eliminated most of the difficulties.

The binomial distribution is illustrated with several examples.

REFERENCES

1. Eves, Howard. (1969). *Introduction to the History of Mathematics*, 3rd Edition, New York: Holt, Rinehart and Winston.
2. Fitzgerald, Edward J. (1959). *Rubaiyat of Omar Khayyam*, Reprinted in *The Romance of the Rubaiyat*, with introduction and notes by A. J. Arberry, London: Allen & Unwin.
3. Gridgeman, N. T. (1968). "Probability and Sex," *The American Statistician, 22* (3), 29–30.
4. Hogben, Lancelot. (1960). *Mathematics in the Making*, Garden City, N. Y.: Doubleday.
5. Kendall, M. G., and W. R. Buckland. (1971). *A Dictionary of Statistical Terms*, Edinburgh: Oliver & Boyd.
6. Larson, H. J. (1969). *Introduction to Probability Theory and Statistical Inference*, New York: Wiley.

EXERCISES

1. Show that the following four binomial expansions of $(a + b)^3$ are equivalent.

a. $\binom{3}{0}a^0b^3 + \binom{3}{1}a^1b^2 + \binom{3}{2}a^2b^1 + \binom{3}{3}a^3b^0$

b. $\binom{3}{0}a^3b^0 + \binom{3}{1}a^2b^1 + \binom{3}{2}a^1b^2 + \binom{3}{3}a^0b^3$

c. $\binom{3}{3}a^0b^3 + \binom{3}{2}a^1b^2 + \binom{3}{1}a^2b^1 + \binom{3}{0}a^3b^0$

d. $\binom{3}{3}a^3b^0 + \binom{3}{2}a^2b^1 + \binom{3}{1}a^1b^2 + \binom{3}{0}a^0b^3$

2. Construct Pascal's triangle to include $n = 2, 3, \ldots, 10$ and identify the rows that correspond to particular values of n.
3. Construct a tree diagram to describe the possible sequences of outcomes for four independent trials. Identify the branches corresponding to zero successes, three successes, and four successes.
4. A coin is tossed 10 times. Use the binomial formula to find the probability of exactly five heads. Use Table A.2 to calculate the probability of at least eight heads.
5. A student takes an examination for which he has not prepared. He guesses at all of the 10 items on a true-false question. What is the probability of getting at least eight correct answers?
6. A die is thrown six times. What is the probability of exactly one six?
7. Two dice are thrown six times. What is the probability of exactly one pair of sixes?
8. Ten percent of the books in a certain library have never been borrowed. If you select five books at random from the library, what is the probability that at least two of them will not have been borrowed previously?
9. Ten repetitions were made of an experiment in which a clairvoyant tries to guess the color of each card dealt (as described in Lecture 9). The numbers of cards correctly identified are 8, 6, 4, 6, 2, 6, 4, 6, 8, and 0. Make probability calculations to judge whether or not these results support the idea of mental telepathy.
10. It is well known that among rectangles with given perimeters, the square has the largest area. Use this result to argue that the binomial variance is largest when $p = \frac{1}{2}$. Why is it intuitively reasonable that the greatest variability in a binomial variable occurs when $p = \frac{1}{2}$?

11. Graph the binomial variance versus p for $n = 10$. Note that the maximum variance of $n/4$ occurs at $p = \frac{1}{2}$.
12. Given a family of four children, list all of the possible sequences by sex. What is the most probable number of girls? Given that there are two boys and two girls, what is the probability of two boys followed by two girls?
13. An examination includes 10 multiple-choice questions with three possible choices for each answer. If there is only one correct answer for each question, what is the probability of correctly answering at least 90% of the 10 questions by guessing?
14. A gardener plants six hills of squash, six seeds to a hill. If the germination rate of the seed is .95, what is the probability of at least one squash plant in all six hills?
15. A mail-order house ships unassembled bicycles. The shipping inspector makes one last inspection to make sure that all parts have been included. If the probability of a complete shipment is .99, what is the probability that all of the 1000 bicycles shipped in 1 month were shipped complete?
16. One thousand components are connected in series. What reliability is required of each of the components if the reliability required for the system is .99?
17. A woman zealously enters every possible sweepstakes contest. If the chance of winning any given sweepstakes is no greater than 1 in 1 million, what is the probability of winning at least one sweepstakes if she enters 1000 times?
18. With $n = 10$ and $p = .5$, find:
    a. $P(\mu - 3\sigma \leq X \leq \mu + 3\sigma)$.
    b. $P(\mu - 2\sigma \leq X \leq \mu + 2\sigma)$.
    c. $P(\mu - \sigma \leq X \leq \mu + \sigma)$.
19. Verify that the mean and variance of X/n is p and $p(1 - p)/n$.
20. With $n = 30$, find the probability that X/n differs from p by no more than $2\sqrt{p(1 - p)}/\sqrt{n}$ for $p = .1, .2, \ldots, .9$.
21. A random sample of students is taken to study student attitudes toward a change in dormitory rules. It is desired that the estimate favoring the change, X/n, differ from the true fraction, p, by no more than .05. How large must the sample size be in order to give this margin of error with .95 probability?

GRESHAM COLLEGE. The building where Karl Pearson first lectured on probability and statistics. Reproduced by permission from the Guildhall Library, City of London.

11

THE NORMAL DISTRIBUTION

11.1 The **normal distribution**, which we encountered briefly in the lectures on political arithmetic, the casino, and frequency distributions, is the most used probability distribution. It is fitting that there be considerable interest in tracing its origins. Because the distribution is often called the **Gaussian** distribution, it might be supposed that it is due to Gauss (1777–1855). However, the distribution predates the birthdate of Gauss. In giving a date for the first publication of the normal distribution, Walker [7, p. 4] is quite emphatic about November 12, 1733.

11.2 In the last half of the seventeenth century Bernoulli (1654–1705) began work on his *Ars Conjectandi* (*The Art of Conjecture*). This monumental four-part work was published in 1713, 8 years after Bernoulli's death. In Part 4, Bernoulli was concerned with how certainly one can determine a probability of heads for a biased coin by many tosses. Bernoulli showed that the probability can be determined as precisely as desired if enough trials are made. This was an elementary form of the **law of large numbers**; that is, x/n approaches p (in a way that can be made precise) as n approaches infinity.

11.3 While working on the problem posed by Bernoulli, De Moivre (1667–1754) discovered an approximate formula for binomial probabilities. This formula seems to be the first appearance of the normal distribution. According to Walker, the formula was probably discovered in 1721 but was first committed to writing by De Moivre in a seven-page Latin treatise dated November 12, 1733 entitled "Approximatio ad Summam Terminorum Binomii $\overline{a + b}^n$ in Seriem expansi." It was published with the second and third editions (1738, 1756) of *The Doctrine of Chances* and with certain copies of *Miscellanea Analytica* (1730). Pearson [6] discovered a copy of the pamphlet in the library of University College London, and called attention to De Moivre as the apparent father of the normal distribution. Adams [1] recognizes that De Moivre obtained the normal distribution but doubts that De Moivre recognized the normal distribution as a distribution in its own right.

11.4 The statistical use of the normal distribution began with the mathematical astronomers Laplace (1749–1827) and Gauss (1777–1855). The first tables giving probabilities (areas under the curve) were published in 1799 by the French physicist Kramp; it was first used for description of data outside the physical sciences by the Belgian statistician Quetelet (1796–1874). The distribution was first called **normal** by Pearson in 1893 and is best known by this name today.

11.5 What was this distribution that so captured the imagination of scientists in the late eighteenth and nineteenth centuries? Galton [3] was so impressed by it that he wrote: "I know of scarce anything so apt to impress the imagination as the wonderful form of cosmic order expressed by the 'Law of Frequency of Error.'"

The normal distribution is simply a bell-shaped curve (or a collection of bell-shaped curves) that does a remarkable job of fitting the histograms of data from widely divergent sources. Actually, the typical normal curve is sharper at the top than most bells, but it is commonly called bell-shaped, and we will persist in this description. The equation of the curve is given by

$$f(x) = \frac{1}{\sigma\sqrt{2\pi}} e^{-(x-\mu)^2/2\sigma^2}$$

where x ranges between $-\infty$ and $+\infty$ and μ represents the mean and σ^2, the variance. The effect of varying μ and σ is displayed in Figure 11.1. As the mean increases, the entire curve moves to the right. As the variance increases, the curve becomes flatter, with a larger spread.

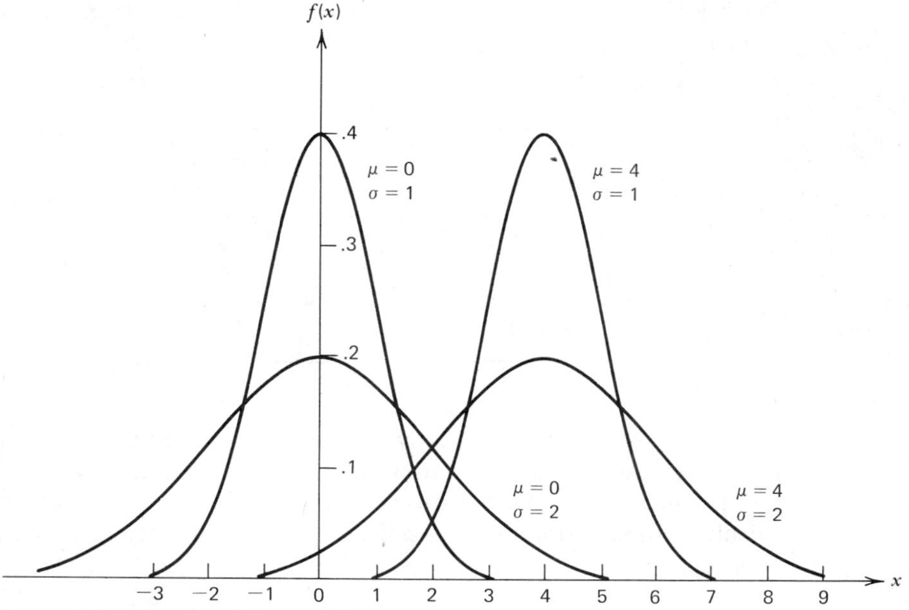

Figure 11.1 Family of Normal Curves.

11.6 A recent television news item concerned a 10-year-old boy who had scored 166 on an IQ test and the efforts of his parents to find special education for him. What is the meaning of such a high IQ score? IQ scores are generally thought to be normally distributed, although the expression IQ is better known to the general public than the normal distribution. Cohen [2] traces the development of mental testing from the ideas of Galton in the 1890s to lawsuits concerning IQ in the 1970s. Another excellent description is given in the Life Science Library volume called *The Mind* [8]. The term **intelligence quotient**, abbreviated IQ, was coined by an American psychologist named Lewis Terman. Originally the idea was to try to obtain the ratio of mental age to chronological age; hence the word quotient. Although the IQ score is no longer determined in that way, the term has lasted.

There are actually many intelligence and aptitude tests in use; scores from most of these are considered to be normally distributed with specified means and variances. Lyman [5] gives a discussion and a conversion table for a number of test scores, including the AGCT (Army General Classification Test) score, the CEEB (College Entrance Examination Board) score, the Stanford-Binet IQ score, and the

Wechsler IQ score. All of these are normally distributed, with means and standard deviations as follows.

Score	Mean	Standard Deviation
AGCT	100	20
CEEB	500	100
Stanford-Binet IQ	100	16
Wechsler IQ	100	15

The 10-year-old boy scoring 166 on the Wechsler IQ test has a score more than four standard deviations above the mean and is in a select group of individuals. We will now discuss the use of the normal curve to calculate probabilities, and we can then say just how select a score of 166 really is.

11.7 Probabilities are represented by areas under the curve. Thus the probability that X lies between 4 and 6, $P(4 \leq X \leq 6)$, is given by the area under the curve between 4 and 6. The area under the curve at any given point is zero and, therefore, the probability that X equals any given value is **zero**. This fact, which strikes many students as puzzling, is easily explained when we remind ourselves that the normal distribution gives a curve that approximately describes a large population. In a large population resembling the normal curve, the chance of drawing any specific value would be small, if not exactly zero.

Many tabulations have been made of the area under the normal curve. In Table A.3, we present tables due to Pearson and Hartley. These tables give $P(X \leq x)$ when $\mu = 0$ and $\sigma = 1$. The case where $\mu = 0$ and $\sigma = 1$ is called the **standard normal** distribution, and it is customary to denote a standard normal variable by Z. We will adhere to this custom in subsequent sections.

With the use of this table and a little effort, one can find any desired area under the curve. It may be helpful to make sketches showing the area known and the area desired. Then, keeping in mind that the curve is symmetric about zero, we can find the desired area much in the manner of solving the "pie problems" of elementary school.

Figure 11.2 Using Normal Tables.

Example. What is $P(-1.5 \leq Z \leq 1)$? Looking at the sketch in Figure 11.2, we can see that

$$\begin{aligned} P(-1.5 \leq Z \leq 1) &= P(Z \leq 1) - P(Z \leq -1.5) \\ &= P(Z \leq 1) - [1 - P(Z \leq 1.5)] \\ &= .8413 - .0668 \\ &= .7745 \end{aligned}$$

11.8 The real utility of the standard normal table lies in the fact that the same table can be used to find the area under any normal curve, **regardless** of μ and σ. The secret lies in first **standardizing** the variable. This is accomplished by **subtracting the mean** and then **dividing by the standard deviation**. In symbols, this means that $(X - \mu)/\sigma$ is a standard normal variable and can be denoted by Z. If we have mastered the use of the standard normal table, we can then find areas under the normal curve for X by standardizing X. Suppose we want

$$P(a \leq X \leq b)$$

Subtracting μ from each term and dividing by σ, the desired probability becomes

$$P\left(\frac{a-\mu}{\sigma} \leq \frac{X-\mu}{\sigma} \leq \frac{b-\mu}{\sigma}\right)$$

$$P\left(\frac{a-\mu}{\sigma} \leq Z \leq \frac{b-\mu}{\sigma}\right)$$

Example. If scores on a standardized test are normally distributed with mean 100 and standard deviation 15, what fraction of scores will lie between 85 and 130? The answer is given by

$$P(85 \leq X \leq 130)$$
$$= P\left(\frac{85 - 100}{15} \leq Z \leq \frac{130 - 100}{15}\right)$$
$$= P(-1 \leq Z \leq 2)$$
$$= .9772 - .1587$$
$$= .8185$$

A graphical approach to this solution is given by the sketch in Figure 11.3.

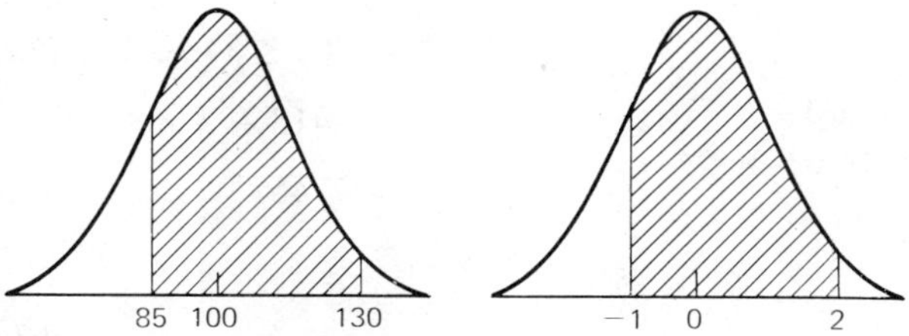

Figure 11.3 Fraction of Standardized Scores.

It is essential that students learn to use the standard normal table. We suggest that the student use sketches such as those in Figures 11.2 and 11.3.

Returning to the case of the 10-year-old with an IQ of 166, we might ask what fraction of the population has a higher IQ. That is, we desire

$$P(X \geq 166)$$

Standardizing the score by subtracting $\mu = 100$ and dividing by $\sigma = 15$, we obtain

$$P(Z \geq 4.4)$$

From Table A.3 we find this probability to be less than .00003. Thus there are very few individuals with higher IQs.

11.9 The normal distribution provides an adequate description, not only of IQ scores, but also of data from a wide range of origins. Many sets of data obtained by measurement processes are normally distributed.

SUMMARY. The **normal distribution** was popularized by Gauss in the early nineteenth century and was generally known as the **Gaussian distribution**. Pearson first called it the **normal distribution** and pointed to De Moivre as the originator. Still, it is often called the Gaussian distribution.

This distribution of a continuous variable is a symmetric, somewhat bell-shaped curve with equation

$$f(x) = \frac{1}{\sigma\sqrt{2\pi}}\, e^{-(x-\mu)^2/2\sigma^2}$$

The population constants, or **parameters**, μ and σ^2, completely specify the distribution.

The normal distribution has been extensively tabled for the **standard normal** case where the **mean** is **zero** and the **variance** is **unity**. This allows calculation of probabilities for any normal distribution through standardizing the variable. **Standardizing** consists of first **subtracting the mean** and then dividing by the **standard deviation.**

IQ scores and scores on many **standardized tests** are generally regarded as having a normal distribution. In addition, **grading on the curve** assumes a normal distribution of test scores.

REFERENCES

1. Adams, William J. (1974). *The Life and Times of the Central Limit Theorem*, New York : Kaedmon.
2. Cohen, Daniel. (1974). *Intelligence, What Is It?* New York : M. Evans.
3. Galton, Francis. (1888). *Natural Inheritance*, London: Macmillan.
4. Johnson, N. L., and S. Kotz. (1970). *Continuous Univariate Distributions*, Vol. 1, Boston : Houghton Mifflin.
5. Lyman, Howard B. (1971). *Test Scores and What They Mean*, Englewood Cliffs, N.J. : Prentice-Hall.
6. Pearson, Karl. (1924). "Historical Note on the Origin of the Normal Curve of Errors," *Biometrika, 16*, 402–404.
7. Walker, Helen M. (1931). *Studies in the History of Statistical Method*, Baltimore : Williams & Wilkins.
8. Wilson, John Rowan, and the editors of *Life* (1964). *The Mind*, New York : Time Inc. (Life Science Library).

EXERCISES

1. If five observations are obtained at random from a normal distribution, what is the probability that at least two of the observations will be greater than μ?
2. Using Table A.3, find:
   a. $P(Z > 1.5)$.
   b. $P(Z > 1.64)$.
   c. $P(Z > 1.645)$.
   d. $P(-1.96 < Z < 1.96)$.
   e. $P(-1.645 < Z < 1.645)$.
3. If $\mu = 100$ and $\sigma = 20$, find $P(X > 120)$.
4. If $\mu = 500$ and $\sigma = 100$, find $P(350 < X < 800)$.
5. An instructor wishes to "grade on the curve." The students' scores for a multiple-section course seem to be normally distributed with mean 70 and a standard deviation of 8. If the instructor wishes to give 10% As, what should be the dividing line between an A grade and a B grade?
6. Although the table does not include such large values, about what is the probability that Z lies between 6 and 9?
7. If the scores on an intelligence test were normally distributed with mean 100 and a standard deviation of 16, how high would a person have to score to receive a genius rating if we give such ratings to only 0.5% of the people taking the exam?
8. Using Table A.3, graph z on the horizontal axis versus the cumulative probability on the vertical axis.
9. If a student's standardized CEEB score was 1.2, what was the raw score?
10. If $\mu = 120$ and $\sigma = 15$, find a constant c so that $P(-c \leq X \leq c) = .5$.
11. If $\mu = -150$ and $\sigma = 15$, find a constant c so that $P(-c \leq X \leq c) = .75$.
12. The percent of alcohol in a nasal decongestant syrup is advertised to be 1.4%. The manufacturing process is such that the percent alcohol is normal with mean 1.4% and standard deviation 0.1%. Find the probability that the percent alcohol in one batch exceeds 1.5%.
13. A breakfast cereal is packaged and sold in 14-ounce packages. Detailed analyses by the shipping inspector reveals that the actual weight is a normal random variable with standard deviation 0.2 ounces. If the fraction of packages with weight less than or equal to 14.5 ounces is .9, what is the mean weight of 14-ounce packages?

14. A person is told that he is overweight because for his height he should weigh between 132 and 145 pounds. If weight is a normal variable and 95% of people his height have weights within the prescribed range, about what is the average weight?
15. An electrical system has 10 components connected in series. Each component will function properly as long as the voltage of one of its components lies between the engineering specifications of 100 and 130 volts. The voltage of each device is a normal variable with a mean of 115 volts and a standard deviation of 5 volts. What is the reliability for the system?
16. The number of monthly orders received by a wholesale house for a particular item seems to be a binomial random variable with $n = 100{,}000$ *and* $p = .4$. The wholesale price for the item is \$15. Calculate the mean and variance for the total monthly sales. Approximate the probability that the total monthly sales exceed \$650,000 by treating the total monthly sales as a normal variable.
17. Plot the binomial probabilities for $n = 10$, $p = .50$. Then represent the binomial distribution by a histogram using intervals -0.5 to 0.5, 0.5 to 1.5, etc.
18. Approximate the binomial probability that $X = 4$ for Exercise 17 by the area under a normal curve from 3.5 to 4.5. Use a mean of $5(np)$ and a variance of $2.5[np(1 - p)]$.

SIMEON DENIS POISSON. 1781–1840. Courtesy Columbia University Library, D. E. Smith Collection.

12

POISSON'S EXPONENTIAL LIMIT

12.1 Poisson (1781–1840), examiner and professor at the *École Polytechnique* of Paris for nearly 40 years, wrote over 300 papers about mathematics, physics, and astronomy. Although his most famous work was *Traité de Mécanique* (1811), he is best known in probability and statistics for the **Poisson distribution**, which was given as a limit of the binomial distribution in 1837. Poisson's exponential limit of the binomial [6] seems to have been predated by the same result obtained by De Moivre. David [2] points out that De Moivre first gave the binomial exponential limit in the first edition of *The Doctrine of Chances* in 1718. Nevertheless, the distribution continues to be known as the Poisson distribution.

12.2 With a rudimentary knowledge of **combinations, limits**, and the **definition of *e*** we can obtain the same result as De Moivre and Poisson. In order to appreciate the Poisson as well as the normal distribution, it is necessary to have a basic understanding of the important mathematical constant, *e*. The definition generally given for *e* is

$$e = \lim_{n \to \infty} \left(1 + \frac{1}{n}\right)^n$$

Fortunately, it is very easy to observe the limiting process called for in this definition. Let us evaluate a few terms in the limiting sequence.

$$(1 + \tfrac{1}{1})^1 = 2.00000$$
$$(1 + \tfrac{1}{2})^2 = 2.25000$$
$$(1 + \tfrac{1}{3})^3 = 2.37037$$
$$(1 + \tfrac{1}{4})^4 = 2.44141$$
$$(1 + \tfrac{1}{5})^5 = 2.48832$$

By continuing this process, students can evaluate e to any desired precision. To five decimal places, e has the value 2.71828. This evaluation is particularly easy on most calculators.

We now remind students of another result that we need to obtain the Poisson distribution as a limiting case of the binomial.

$$\lim_{n \to \infty} \left(1 + \frac{c}{n}\right)^n = e^c$$

This result can also be verified numerically with a calculator.

12.3 With an understanding of e and e^c we can proceed to obtain the limiting form of the binomial distribution as n becomes large and p becomes small in such a way that the product np is constant. First, write the binomial probability function as

$$f(x) = \frac{n!}{x!(n-x)!} p^x(1-p)^{n-x}$$

Next, observing that

$$\begin{aligned} n! &= n(n-1)(n-2)\ldots(n-x+1)(n-x)\ldots(2)(1) \\ &= n(n-1)(n-2)\ldots(n-x+1)(n-x)! \end{aligned}$$

we have that

$$f(x) = \frac{n(n-1)(n-2)\ldots(n-x+1)}{x!} p^x(1-p)^{n-x}$$

Next replace p by λ/n (because $np = \lambda$) to obtain

$$\begin{aligned} f(x) &= \frac{n(n-1)(n-2)\ldots(n-x+1)}{x!} \left(\frac{\lambda}{n}\right)^x \left(1 - \frac{\lambda}{n}\right)^{n-x} \\ &= \frac{n(n-1)(n-2)\ldots(n-x+1)}{n(n)(n)\ldots(n)} \frac{\lambda^x}{x!} \left(1 - \frac{\lambda}{n}\right)^{n-x} \end{aligned}$$

Now consider what happens when n becomes indefinitely large. The first terms approach unity, and we are left with

$$\lim_{n \to \infty} f(x) = \frac{\lambda^x}{x!} \lim_{n \to \infty} \left(1 - \frac{\lambda}{n}\right)^{n-x}$$

Now, as n approaches infinity, so does $n - x$, so

$$\lim_{n\to\infty}\left(1 - \frac{\lambda}{n}\right)^{n-x} = \lim_{n\to\infty}\left(1 - \frac{\lambda}{n}\right)^{n} = e^{-\lambda}$$

Finally, we obtain the famous result that the limit of the binomial is given by the exponential expression

$$f(x) = \frac{\lambda^x e^{-\lambda}}{x!}, \qquad x = 0, 1, 2, \ldots$$

This distribution is known as the Poisson distribution. Naturally, since n was allowed to become infinite, the possible values of x extend from zero to infinity.

12.4 The **moments** of the Poisson are particularly easy to obtain and are well known. The remarkable fact is that both the **mean** and the **variance** are equal to the constant (or parameter) λ. To be explicit,

$$\mu = \lambda$$

and

$$\sigma^2 = \lambda$$

This result can be given intuitive justification as a limiting property of the binomial mean and variance. With n getting large in such a way that np is constant, p gets small as n becomes large. Then

$$\lim_{n\to\infty} np = \lambda$$

and

$$\lim_{n\to\infty} npq = \lim_{p\to 0} \lambda(1 - p)$$
$$= \lambda$$

12.5 The Poisson distribution has been extensively **tabled**, and Table A.4 gives $f(x)$ for x from 0 to 24. For example, if $\lambda = 3.8$, $f(3) = 0.2046$. If $\lambda = 9.0$, $f(3) = 0.0150$.

12.6 One of the classic, if today somewhat quaint, examples of using the Poisson distribution was given in 1898 by Bortkiewicz [1]. He tabulated the number of cavalrymen who died from a **kick** from a horse. Some of the Bortkiewicz data were presented earlier in Lecture 5 (see Fisz [3]).

Taking 10 cavalry corps over 20 years gives a total of 200 annual records. Consider the number of horsekick deaths (r) reported in each of these 200 records and the frequency for each value of r reported. Bortkiewicz's results are given in Table 12.1.

In 109 of the 200 records, no horsekick deaths were reported. In 65 of the 200 records, one horsekick death was reported, etc.

From the data the sample mean is obtained as

$$\frac{(0)(109) + (1)(65) + (2)(22) + (3)(3) + (4)(1)}{200} = 0.61$$

The mean obtained from the data is then used as the value for λ in the probability distribution. This gives us the equation

$$f(x) = \frac{e^{-0.61}(0.61)^x}{x!}$$

which was introduced in Lecture 5 and which can be used to calculate the theoretical probabilities. The values can also be approximated by using Table A.4 with $\lambda = 0.6$.

The Bortkiewicz data is well described by the Poisson distribution. Our interest in this particular example is now purely historical. However, it is easy to conceive of the Poisson distribution arising in this example as a **limiting approximation** to a binomial. A cavalryman will either be killed by horsekick during a year or not. It is reasonable to suppose that the chance of this rare event is the same for all soldiers and that soldiers have independent chances of being killed. Thus the number

Table 12.1 Frequency Distribution of Number of Cavalrymen Killed by Horsekick

r	Frequency	Relative Frequency	Theoretical Probability
0	109	0.545	.544
1	65	0.325	.331
2	22	0.110	.101
3	3	0.015	.021
4	1	0.005	.003

of cavalrymen killed during a year is a binomial variable. However, the probability of being killed, p, is very small and the number of trials, the number of cavalrymen, is very large. Therefore the Poisson limit gives a very good description of these data.

12.7 Another example of using the Poisson distribution for the description of real data was published in 1907 by "Student" [7]. He also gives a derivation of the exponential limit to the binomial without reference to either Poisson or De Moivre. "Student" gives the frequencies of yeast cells counted in a thin layer of liquid spread out over a slide. The frequency counts for 400 small areas are given by Table 12.2.

As with the horsekick data, we can obtain the sample mean $\bar{x} = 0.6825$. Using this value for λ, we can obtain the theoretical probabilities and see that the Poisson distribution does a good job of describing the data. We can also approximate these probabilities by using Table A.4 with $\lambda = 0.7$. Students are encouraged to read this paper to see how the Poisson distribution arises as a limiting form of the binomial.

12.8 Mullet [5] used the Poisson distribution to describe the number of goals scored for and against each of the teams in the National Hockey League in the 1973–1974 season. For example, Boston scored 4.95 goals on the average in its home games. The frequency distribution of number of goals per game is shown in Table 12.3 along with theoretical probabilities calculated by using $\lambda = 4.95$. These probabilities can be closely

Table 12.2 "Student's" Yeast Cell Data

r	Frequency	Relative Frequency	Theoretical Probability
0	213	0.5325	.505
1	128	0.32	.345
2	37	0.0925	.1175
3	18	0.045	.0275
4	3	0.0075	.0046
5	1	0.0025	.0006

Source. Reproduced from *Biometrika, 5* "Student," On the error of counting with a haemacytometer. Copyright © 1906 by the Biometrika Trustees. Reprinted by permission of the Biometrika Trustees.

Table 12.3 Boston's Home Goals: 1973–1974

r	Frequency	Relative Frequency	Theoretical Probability
0–1	1	0.0256	.0421
2	2	0.0513	.0868
3	5	0.1282	.1432
4	9	0.2308	.1772
5	10	0.2564	.1754
6	5	0.1282	.1447
7	2	0.0513	.1023
8	3	0.0769	.0633
9	1	0.0256	.0348
≥10	1	0.0256	.0302

Source. Reproduced from *The American Statistician, 31,* Gary M. Mullet, Simeon Poisson, and the National Hockey League.

approximated by using Table A.4 with $\lambda = 5.0$. Mullet suggests that because each goal is like a random observation from a Poisson distribution, the game of hockey is quite predictable.

12.9 In summary, the Poisson distribution can be used as an approximation for the binomial (large n and small p) and as a distribution in its own right for counting data of certain types, such as number of radioactive particles emitted in a given time period, number of telephone calls received in a given time period, number of equipment failures in a given time period, and even number of cavalrymen killed by horsekick in a year.

SUMMARY. The **Poisson distribution**, like the normal distribution, was first obtained by De Moivre but generally credited to someone else, in this case Poisson.

With an understanding of the constant e as the limit of $(1 + 1/n)^n$ as n becomes large, it is possible to obtain the **limit** of the **binomial distribution**. As n goes to infinity and p goes to zero so that $np = \lambda$, a constant, the limit of the binomial distribution is

$$f(x) = e^{-\lambda}\lambda^x/x!$$

Both the **mean** and **variance** of the Poisson distribution are found to equal λ. This is a somewhat surprising result, but it can also be verified by taking the limit of the binomial mean and variance as n goes to infinity with $np = \lambda$.

The Poisson distribution is also extensively **tabled** and is used to describe the probability of **rare events** such as horsekick deaths, lightning flashes, and equipment failures.

REFERENCES

1. Bortkiewicz, L. (1898). *Das Gesetz der kleinen Zahlen*, Leipzig: Teubner.
2. David, F. N. (1973). *Chambers's Encyclopedia*, p. 146, London: International Learning Systems Limited.
3. Fisz, Marek. (1963). *Probability Theory and Mathematical Statistics*, New York: Wiley.
4. Johnson, N. L., and Kotz, S. (1969). *Distributions in Statistics: Discrete Distributions*, Boston: Houghton Mifflin.
5. Mullet, Gary M. (1977). "Simeon Poisson and the National Hockey League," *The American Statistician, 31*, 8–12.
6. Poisson, Simeon D. (1837). *Recherches sur la Probabilité des Jugements on Matière Criminelle et en Matière Civile.*
7. "Student." (1906). "On the Error of Counting with a Haemacytometer," *Biometrika*, 5, 351–360.

EXERCISES

1. Using a calculator, evaluate $[1 + (1/n)]^n$ for $n = 6, 7, 8, 9, 10$. Does the sequence seem to be converging toward the stated value?
2. Using a calculator with a key for e, evaluate e^3. Next calculate $[1 + (3/n)]^n$ for $n = 1, 2, 3, \ldots, 10$. Does the sequence seem to be converging toward the value for e^3?
3. For a binomial distribution with $n = 50$ and $p = .1$, calculate the probabilities for $x = 0, 5, 10, 15, 20$. Using the Poisson formula with $\lambda = np = 5$, calculate the probability that $x = 0, 5, 10, 15, 20$. Compare the Poisson probabilities with the binomial probabilities.
4. Using Table A.4, calculate the probability of at least five occurrences for $\lambda = 1, 2, 3, 4, 5$.

5. Using the value of 0.61 for λ, calculate the "theoretical probabilities" for the Bortkiewicz data. Graph the observed relative frequencies and theoretical probabilities on the same graph paper.
6. Using the value of 0.6825 for λ, calculate the theoretical probabilities for "Student's" yeast cell data. Graph the observed relative frequencies and theoretical probabilities on the same graph paper.
7. Find a set of data that should be well described by the Poisson distribution (duration of strikes, numbers of traffic accidents, bacterial counts, numbers of severe storms, etc.). Using the average of the data as the value for λ, calculate the theoretical frequencies and compare with the observed.
8. Using the value of 4.95 for λ calculate the theoretical probabilities for the hockey goal data of Table 12.3. Graph the observed relative frequencies and theoretical probabilities on the same graph.
9. Graph the Poisson probability function for $\lambda = 1, 2, \ldots, 10$.
10. Represent the Poisson distribution with $\lambda = 10$ by a histogram with classes -0.5 to 0.5, 0.5 to 1.5, 1.5 to 2.5, etc. Approximate $P(X = 8)$ (the area of the bar centered on 8) by the area under a normal curve with mean $10(\lambda)$ and variance $10(\lambda)$.
11. Use the Poisson distribution to calculate the binomial probability of Lecture 10, Exercise 15, with $\lambda = np$.
12. Use the Poisson distribution to calculate the binomial probability of Lecture 10, Exercise 17.
13. A municipal power plant has experienced occasional power outages for some parts of town. A new industry is considering locating a new facility in the town but is concerned about the possibility of a power shortage. The number of power outages per year seems to be a Poisson with $\lambda = 6$. What is the probability of no more than two power failures per year?
14. The manager of a large pecan orchard uses a spray program to control the pecan weevil population. The number of pecan weevils per tree is a Poisson variable with parameter λ. He wishes to control the weevil population so that the probability is .95 that there are no more than 50 weevils per tree. What should be the mean number of weevils per tree?
15. The physical plant manager for a university has become worried about the saturation of university telephone lines and is considering installing some lines for emergency calls only. If the number of calls reaches 20,000 calls per hour, the telephone system for offices is

saturated. What is the greatest that λ can be so that the probability is .95 that the number of calls does not exceed 20,000? Use the normal approximation to determine an answer.

16. Equipment problems with a fleet of airplanes necessitates a new maintenance schedule. Any airplane with more than three minor difficulties in 250 hours of operation receives a complete maintenance check. The number of equipment problems per 1000 hours of operation is a Poisson variable with parameter $\lambda = 10$. What fraction of planes in the fleet will be expected to have complete maintenance checks after 250 hours of operation?
17. A motorist has received her third traffic ticket in a month for running a red light at an intersection on the way to work and has become convinced that the traffic light is not functioning properly. She conducts her own study of the traffic light and determines that the number of cars it allows through on green is a Poisson variable with parameter $\lambda = 3$. What is the probability of a traffic violation for the first car waiting? For the second car? For the third car?

LARGE AND SMALL SAMPLES

SURVEY SAMPLING. Always a chance for an error. UPI.

13

SAMPLING DISTRIBUTIONS AND SURVEY SAMPLING

13.1 *For all sad words of tongue or pen*
The saddest are these: "It might have been!"

John Greenleaf Whittier

In Lecture 1 numerous references were made to early **censuses** and **census taking**. A modern development in census taking is the deliberate use of **sampling**. In the 1970 U.S. census of population and housing most of the information was collected by sampling (see *Procedural History, 1970 Census of Population and Housing*). The complete census was used for the population items of age, sex, race, relationship to head of household, and marital status. For most of the other information on population, the basic sample was a 20% sample; a smaller sample was taken on some items.

We are exposed almost daily to the results of **public opinion polls** on every item under the sun. Such polls have wide acceptance by the general public, although a healthy scepticism remains—perhaps because of disasters such as the polls that predicted Landon's victory over Roosevelt in 1936 and Dewey's victory over Truman in 1948. Despite general acceptance, there is a mystique associated with drawing valid conclusions from a sample. We wish to try to remove some of the mystery; to do so, we will consider in detail a small numerical example.

13.2 Consider the miniature city consisting of seven households, as described in Table 13.1. From this table we can easily calculate the **average annual household income**, μ, to be \$29,000.

Table 13.1 Miniature City

Households	Annual Income (thousands of dollars)
North of river	
1	30
2	35
3	78
South of river	
4	15
5	13
6	12
7	20

Suppose that Table 13.1 were not available to us and we decided to find out about annual income by taking a **simple random sample** of size 5. If the sample consisted of households 1, 2, 4, 6, and 7, the **sample average,** $\bar{x}$, would be \$22,400. In reporting this value we would realize that other values of $\bar{x}$ could have occurred, and we would be led to describing the uncertainty in the single value of $\bar{x}$ we obtained in terms of the values of $\bar{x}$ that **might have been.**

The careful consideration of what might have been is an important part of statistical thinking. We must consider the meaning of a statistic calculated from a single sample from a given population. We consider the conceptual population of statistic values that could have been generated if all possible samples had been drawn. For this conceptual population there is a probability distribution called a **sampling distribution**, and this distribution is then used to interpret the meaning of a statistic value from a single sample.

13.3 With the population of Table 13.1, what are the values of $\bar{x}$ that could have been generated by simple random sampling? With this small population we can actually list all possible samples, with the resulting $\bar{x}$ as in Table 13.2.

From Table 13.2 we can calculate the mean and variance of $\bar{x}$ to be

$$\mu_{\bar{x}} = 29$$

and

$$\sigma_{\bar{x}}^2 = \Sigma(\bar{x} - 29)^2/21$$
$$= 31.05$$

Table 13.2 Simple Random Sampling

Sample Households	$\bar{x}$	Probability
1 2 3 4 5	34.2	$\frac{1}{21}$
1 2 3 4 6	34.0	$\frac{1}{21}$
1 2 3 5 6	33.6	$\frac{1}{21}$
1 2 4 5 6	21.0	$\frac{1}{21}$
1 3 4 5 6	29.6	$\frac{1}{21}$
2 3 4 5 6	30.6	$\frac{1}{21}$
1 2 3 4 7	35.6	$\frac{1}{21}$
1 2 3 5 7	35.2	$\frac{1}{21}$
1 2 4 5 7	22.6	$\frac{1}{21}$
1 3 4 5 7	31.2	$\frac{1}{21}$
2 3 4 5 7	32.2	$\frac{1}{21}$
1 2 3 6 7	35.0	$\frac{1}{21}$
1 2 4 6 7	22.4	$\frac{1}{21}$
1 3 4 6 7	31.0	$\frac{1}{21}$
2 3 4 6 7	32.0	$\frac{1}{21}$
1 2 5 6 7	22.0	$\frac{1}{21}$
1 3 5 6 7	30.6	$\frac{1}{21}$
2 3 5 6 7	31.6	$\frac{1}{21}$
1 4 5 6 7	18.0	$\frac{1}{21}$
2 4 5 6 7	19.0	$\frac{1}{21}$
3 4 5 6 7	27.6	$\frac{1}{21}$

Table 13.2 provides a fundamental way of thinking about all possible samples, but it is not necessary to construct the table in order to get the variance of $\bar{x}$. In the theory of survey sampling this variance is derived to be

$$\sigma_{\bar{x}}^2 = \frac{N-n}{N} \cdot \frac{S^2}{n}$$

where

N = the population size
n = the sample size
S^2 = the population total mean square $\Sigma(x-\mu)^2/(N-1)$

From the data in Table 13.1 we find that

$$S^2 = 543.33$$

so that

$$\frac{N-n}{N}\frac{S^2}{n} = \frac{2}{7}\frac{S^2}{5}$$

$$= 31.05$$

Let us summarize what we have learned. If we take a simple random sample of size n and calculate the sample average, $\bar{x}$, we can make the following statements.

1. The average of all possible $\bar{x}$s that could have been obtained is the mean of the original population, μ. In terms of expected value, we say $E(\bar{x}) = \mu$ and call $\bar{x}$ **unbiased**.
2. The variance of all possible $\bar{x}$s is given by the formula

$$\sigma_{\bar{x}}^2 = \frac{N-n}{N}\frac{S^2}{n}$$

For n small relative to N, the variance of $\bar{x}$ is closely approximated by

$$\sigma_{\bar{x}}^2 = \frac{\sigma^2}{n}$$

13.4 The variance formula involves unknown **parameters** and, therefore, must be estimated. An unbiased estimate of the variance is given by

$$s_{\bar{x}}^2 = \frac{N-n}{N}\frac{s^2}{n}$$

where $s^2 = \Sigma(x_i - \bar{x})^2/(n-1)$, the sample total mean square. To illustrate, suppose we had drawn the sample $\{1, 3, 4, 5, 6\}$. Then $s^2 = 785.3$ so

$$s_{\bar{x}}^2 = \frac{7-5}{7}\frac{785.3}{5} = 44.87$$

and $s_{\bar{x}} = 6.70$.

If the population mean income were unknown to us we would then report our sample mean of $\bar{x} = 29.6$ with an estimated standard deviation of 6.70.

13.5 Quite frequently the population is blocked (or **stratified**) into two or more blocks (**strata**) and a random sample is taken from each stratum. For our miniature city, suppose we stratify our city into two strata:

Table 13.3 Stratified Random Sampling

Households	$\bar{x}_N$	Households	$\bar{x}_S$	$\bar{x}_{st}$
1 2	32.5	4 5 6	13.33	21.55
1 2	32.5	4 5 7	16.00	23.07
1 2	32.5	4 6 7	15.67	22.88
1 2	32.5	5 6 7	15.00	22.50
1 3	54.0	4 5 6	13.33	30.76
1 3	54.0	4 5 7	16.00	32.29
1 3	54.0	4 6 7	15.67	32.09
1 3	54.0	5 6 7	15.00	31.71
2 3	56.5	4 5 6	13.33	31.83
2 3	56.5	4 5 7	16.00	33.36
2 3	56.5	4 6 7	15.67	33.17
2 3	56.5	5 6 7	15.00	32.79

households north of the river and households south of the river. Then we take a random sample of two households north of the river and a random sample of three households south of the river. We can perform the calculations of the previous section for both sides of the river. This gives the two sample averages $\bar{x}_N$ and $\bar{x}_S$, which we combine by calculating $\bar{x}_{st} = (3\bar{x}_N + 4\bar{x}_S)/7$. These calculations are summarized in Table 13.3. The average of the possible values of the stratified estimate is 29.00 as with $\bar{x}$, so $\bar{x}_{st}$ is also an unbiased estimate of μ. A comparison of the values of $\bar{x}_{st}$ in Table 13.3 with the values of $\bar{x}$ in Table 13.2 reveals that the stratified estimate is much less variable. This is borne out by the fact that the variance of $\bar{x}_{st}$ is 21.66, whereas the variance of $\bar{x}$ was 31.05.

13.6 In actually conducting a survey of a city we must devise some way of eliciting the information we desire. In fact, we probably will not have a complete list of households and will have to devise some other approach. If we are using door-to-door interviewers we might use a map of the city. In such cases we will have to make some compromises with random sampling because all dwelling units will not be shown on the map. For example, we may choose blocks at random and take every third residence within blocks. If we conduct a telephone survey, we will have to be satisfied with existing telephone directories, realizing that some people are not listed. We may use a mail-out questionnaire if we have a reasonable list of residences (e.g., the telephone directory).

Regardless of the method we use, there will be major problems not encountered with our miniature city example. Sometimes the **nonresponse** rate will be very high. If we have planned a fairly small sample to begin with, nonresponses may give us such a small sample that our work is not informative. Other problems can arise if our instrument or questionnaire has not been **pretested** to insure that it makes sense to the people being surveyed.

13.7 We have already considered what happens when the population size, N, is large **relative** to the sample size, n. This usually results in simpler formulas, and we are therefore interested in random sampling from so-called "infinite" populations. Consider a very large population with one-third 1s, one-third 2s, and one-third 3s, so that $\mu = 2$ and $\sigma^2 = \frac{2}{3}$. Consider a sample of size $n = 2$ and consider the sampling distribution of $\bar{x}$. The possible samples with values of $\bar{x}$ are given in Table 13.4. From Table 13.4 we can obtain easily the sampling distribution of $\bar{x}$ as given in Table 13.5.

Table 13.4 Sampling an Infinite Population

Sample	x_1	x_2	$\bar{x}$	Probability
1	1	1	1	$\frac{1}{9}$
2	1	2	$\frac{3}{2}$	$\frac{1}{9}$
3	2	1	$\frac{3}{2}$	$\frac{1}{9}$
4	1	3	2	$\frac{1}{9}$
5	3	1	2	$\frac{1}{9}$
6	2	2	2	$\frac{1}{9}$
7	2	3	$\frac{5}{2}$	$\frac{1}{9}$
8	3	2	$\frac{5}{2}$	$\frac{1}{9}$
9	3	3	3	$\frac{1}{9}$

Table 13.5
Sampling Distribution of $\bar{x}$

$\bar{x}$	$P(\bar{x})$
1	$\frac{1}{9}$
$\frac{3}{2}$	$\frac{2}{9}$
2	$\frac{3}{9}$
$\frac{5}{2}$	$\frac{2}{9}$
3	$\frac{1}{9}$

From this table we can calculate the mean and variance of $\bar{x}$ and find that

$$\mu_{\bar{x}} = 2 = \mu$$

$$\sigma_{\bar{x}}^2 = \tfrac{1}{3} = \sigma^2/n$$

13.8 Many ideas concerning **sampling distributions** were published as early as 1832 (Encke [4]) and many results were given in the 1860s (Airy [1]). However, the paper by Pearson [7] is often regarded as the beginning of modern sampling distribution theory. The development of the subject received a great impetus with publication of "Student's" paper in 1908. Beginning with his paper [5] in 1915, Fisher obtained and published the sampling distributions of many statistics being used for the analysis of data. Following the work of Fisher, countless researchers have obtained the sampling distribution of virtually every conceivable statistic.

SUMMARY. In recent years the U.S. Census Bureau has made extensive use of **sampling** to obtain some of the census information. When a simple random sample is taken from a population, the **sample mean**, $\bar{x}$, is used to **estimate** the **population mean**. When a **stratified random sample** is taken, a weighted combination of the strata sample means, $\bar{x}_{st}$, is used. To compare the two estimates requires consideration of the different values that could be obtained from all possible samples. This leads to what is called a **sampling distribution**.

The sampling distributions of both $\bar{x}$ and $\bar{x}_{st}$ have the same mean as the original population. Therefore both $\bar{x}$ and $\bar{x}_{st}$ are **unbiased** estimates of μ. However, if the strata are chosen well, the variance of $\bar{x}_{st}$ will be smaller than the variance of $\bar{x}$.

The variance of the sample mean from a simple random sample is given by

$$\sigma_{\bar{x}}^2 = \frac{N-n}{N-1}\frac{\sigma^2}{n}$$

If the population is quite large, infinite for practical purposes, the variance of $\bar{x}$ is given by σ^2/n.

REFERENCES

1. Airy, G. B. (1861). *Theory of Errors of Observations*, London.
2. Bureau of the Census. (1976). *1970 Census of Population and Housing, Procedural History*, PHC(R)-1.

3. Davis, James A. (1971). *Elementary Survey Analysis*, Englewood Cliffs, N.J.: Prentice-Hall.

4. Encke, Johann Franz. (1832). "Uber die Methode der Kleinsten Quadrate," *Berliner Astronomisches Jahrbuch für 1834.*

5. Fisher, R. A. (1915). "Frequency Distribution of the Values of the Correlation Coefficient in Samples from an Indefinitely Large Population," *Biometrika, 10*, 507–521.

6. Mendenhall, William, Lyman Ott, and Richard L. Scheaffer. (1971). *Elementary Survey Sampling*, Belmont, Calif.: Wadsworth.

7. Pearson, Karl. (1900) On the Criterion that a Given System of Deviations from the Probable in the Case of a Correlated System of Variables Is Such that It Can Be Reasonably Supposed to Have Arisen from Random Sampling," *Philosophical Magazine*, 5(50), 157–175.

EXERCISES

1. Verify the value for the variance of the sample mean from Table 13.2. Show that this same value can be obtained from the formula

$$\sigma_{\bar{x}}^2 = \frac{N - n}{N} \frac{S^2}{n}$$

2. Suppose that in Section 13.2 we had drawn the sample $\{1, 2, 4, 6, 7\}$. Use this sample to estimate μ. Estimate the variance of your estimate of μ.

3. Suppose that we use stratified random sampling, resulting in the choice of $\{1, 3\}$ north of the river and $\{4, 6, 7\}$ south of the river. Estimate the average annual household income.

4. A stratified random sample was conducted to estimate the average weekly food costs of married couples in a small city. The couples were stratified into four strata: (*a*) couples aged 20 to 40 without children, (*b*) couples aged 40 and up without children, (*c*) couples with one or two children, and (*d*) couples with more than two children. The stratum sizes, sample sizes, sample means, and sample variances were as follows.

Stratum	N	n	$\bar{x}$	s^2
1	1000	10	44	64
2	2000	15	39	76
3	3000	25	58	46
4	4000	15	74	48

Estimate the average weekly food cost.

5. Consider a very large population with the population distribution given by:

x	$P(x)$
1	$\frac{1}{6}$
2	$\frac{1}{2}$
3	$\frac{1}{3}$

   a. Calculate the mean μ and variance, σ^2.
   b. For a sample of size 2, obtain the sampling distributions for $\bar{x}$, s^2, and $x_{\max} = \max(x_1, x_2)$.
   c. For each of the distributions in part b, find the mean and variance.
   d. Verify that $\mu_{\bar{x}} = \mu$, $\sigma^2_{\bar{x}} = \sigma^2/n$, and $\mu_{s^2} = \sigma^2$.

6. The owner of a chain of supermarkets is considering building another store in a neighboring town and hires an opinion research firm to conduct a survey of shoppers' attitudes. A random sample of 50 households is selected from the city population of 5000 households. One of the many questions asks the amount of the monthly shopping budget. The total monthly budget reported for the sample is \$10,000, and the s^2 for the sample is \$2000. Estimate:
   a. The average monthly shopping budget for the town.
   b. The variance of the sample mean.
   c. The total monthly shopping budget for the town.
   d. The variance of the estimate in part c.

7. A director of student advisement is making a study of why students withdraw from the university after the freshman year. From the many students who have withdrawn over the past 10 years she selects some at random and obtains 26 who say they dropped because of grade difficulties. The 26 students have an average grade

point average of 2.137 and an s^2 of 0.2638. What is the estimate of the grade point average for those students who dropped out because of grade difficulties? What is your estimate of the variance of this estimate? If the first-year grade point average for those who finish is 2.417, do you think that the data support the reason given for withdrawing?

8. Draw a random sample of two-digit numbers of size 15 from Table A.1. Then calculate the sample mean and estimate the standard deviation of the sample mean. Are the results consistent with a population mean of 49.5 (the mean of the population being simulated by the random number table)?

9. In a population some of the xs are one and the others are zero. If Np of the xs are one and $N(1 - p) = Nq$ are zero, show that

$$\sum_{1}^{N} (x_i - \mu)^2 = \sum_{1}^{N} x_i^2 - N\mu^2 = Np(1 - p)$$

$$= Npq$$

10. A survey is conducted to estimate the fraction of people favoring a certain proposition. Let x for an individual be one if she favors the proposition, zero if she opposes it. Then show that the sample fraction favoring the proposition, $\hat{p}$, is equal to $\bar{x}$. Also show that the variance of p is

$$\frac{N - n}{N - 1} \cdot \frac{pq}{n}$$

ABRAHAM DE MOIVRE. 1667–1754. The Granger Collection.

14

THE CENTRAL LIMIT THEOREM

14.1 There is a theorem considered by many to be the most important theorem in statistics. It is concerned with the distribution of a sum (or mean) of random variables. As the number of random variables increases, the distribution of the sum (or mean) approaches a normal distribution as a limit. It seems natural to call the theorem a limit theorem. Because of its importance or **centrality** to statistics, it seems natural to call it the **central limit theorem**, the name introduced by Polya [3] in 1920.

14.2 As intimated when we discussed the normal distribution, a vast body of statistical methods exists for the analysis of data which are normally distributed. In fact, many sets of observations found in nature are well described by normal distributions. In addition, many analyses are performed upon numbers which are sums (or means) of other observations which may not be normally distributed. Let us cite two examples.

1. In educational research the basic data are often obtained from responses to questionnaires. The response to each item on the questionnaire may be one of 5 possible responses, -2, -1, 0, 1, 2. The item responses are then highly discrete and the distribution of responses nonnormal. However, we may be interested in a total score for each respondent and the total score may be a sum of as many as 50 item responses. For the distribution of total scores, normality is a very plausible assumption.
2. In life testing of electronic devices the time to failure (operating time before failure) of individual devices is recorded. Such failure times typically have nonnormal distributions. However, the total (or

average) time to failure for a sample of several devices will have a much more symmetric distribution and, for moderately large samples, the normality assumption for the total (or mean) will be reasonable.

14.3 Prior to De Moivre [2] there was no normal distribution and, naturally, no explicit recognition of approximate normality. The step that De Moivre made was simultaneously a step toward formulation of the normal distribution and a step toward formulation of the central limit theorem. De Moivre obtained what we would describe in modern terminology as a normal approximation to the binomial.

14.4 As an empirical introduction to the central limit theorem from a viewpoint similar to that of De Moivre, let us consider the distribution of the binomial variable X as we allow n to increase, holding p constant. From Table A.2. we extract the following probabilities for $p = \frac{1}{3}$ (see Table 14.1). To see the effect as n increases from 1 to 5, consider the pictures in Figure 14.1. As n increases from 1 to 5, we see three characteristics of the normal distribution beginning to emerge.

1. Probability is being distributed over many points on the number line.
2. The distribution of probability is becoming more symmetric.
3. The characteristic bell-like shape is hinted at with $n = 3$ and is unmistakeable with $n = 4$ and $n = 5$.

As suggested by this example, the **convergence to normality** as n becomes infinitely large is incredibly rapid, so that with very small n, the distribution is beginning to become normal.

Table 14.1 Binomial Probabilities ($p = \frac{1}{3}$)

x	$n = 1$	$n = 2$	$n = 3$	$n = 4$	$n = 5$
0	0.6667	0.4444	0.2963	0.1975	0.1317
1	0.3333	0.4444	0.4444	0.3951	0.3292
2		0.1111	0.2222	0.2963	0.3292
3			0.0370	0.0988	0.1646
4				0.0123	0.0412
5					0.0041

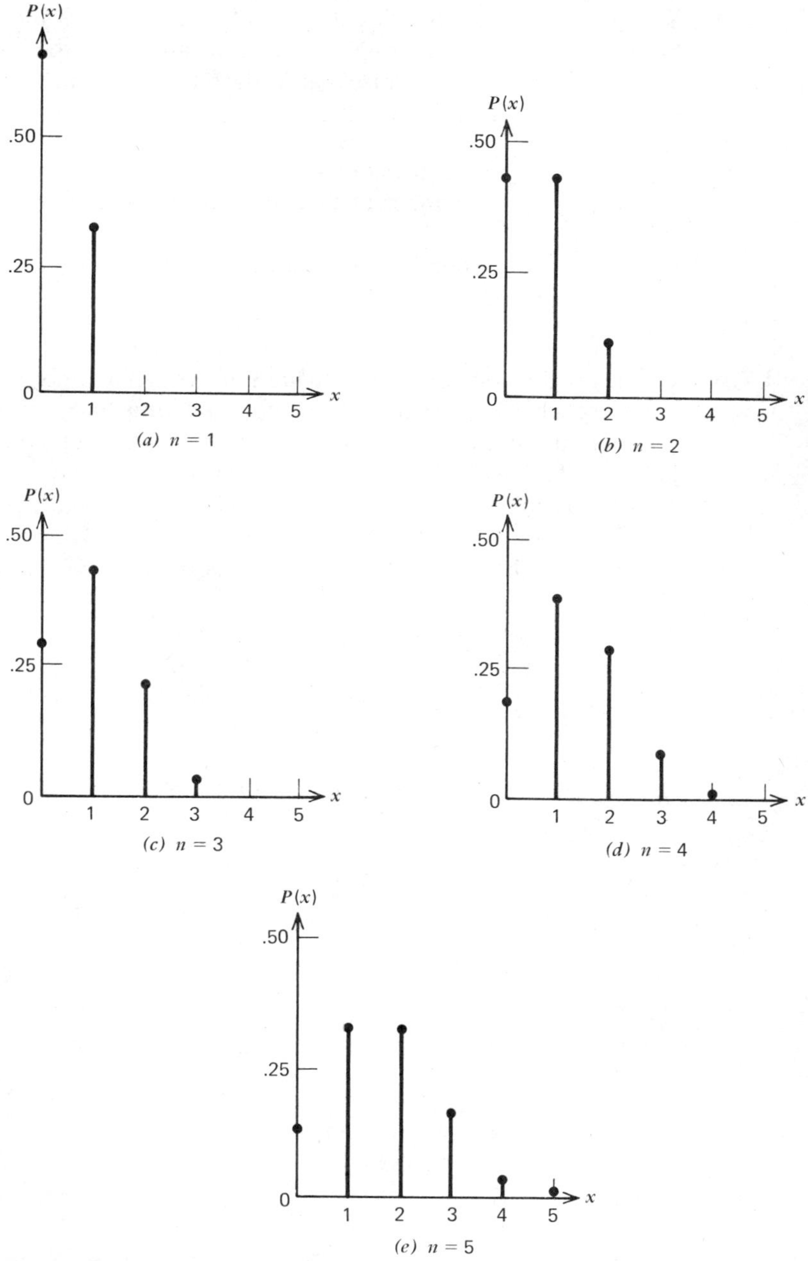

Figure 14.1 Binomial Probabilities ($p = \frac{1}{3}$). (*a*) $n = 1$. (*b*) $n = 2$. (*c*) $n = 3$. (*d*) $n = 4$. (*e*) $n = 5$.

14.5 Why does the binomial distribution of X approach a normal distribution as n becomes infinite? In order to appeal to the central limit theorem, we need to recognize that X, the number of successes in n trials, is the sum of n variables. In fact,

$$\begin{aligned} X = &\text{ number of successes on first trial} \\ &+ \text{number of successes on second trial} \\ &+ \cdots \\ &+ \text{number of successes on } n\text{th trial} \end{aligned}$$

14.6 In order to understand the central limit theorem, it is important to follow the process of sampling from a parent population. Suppose the original population is uniform, as shown in Table 14.2. Next, let us consider the **joint distribution** of X_1 and X_2, the observations in a random sample of size 2. The joint distribution of X_1 and X_2 is shown in Table 14.3. The numbers in parentheses give the sum, $X_1 + X_2$. The probability for any value of the sum is then given by adding the probabilities on the

Table 14.2
Original Population Distribution

x	$P(x)$
1	$\frac{1}{4}$
2	$\frac{1}{4}$
3	$\frac{1}{4}$
4	$\frac{1}{4}$

Table 14.3 Joint Distribution of X_1 and X_2

		$P(x_2)$	$\frac{1}{4}$	$\frac{1}{4}$	$\frac{1}{4}$	$\frac{1}{4}$
$P(x_1)$	x_1	x_2	1	2	3	4
$\frac{1}{4}$	1		$\frac{1}{16}$(2)	$\frac{1}{16}$(3)	$\frac{1}{16}$(4)	$\frac{1}{16}$(5)
$\frac{1}{4}$	2		$\frac{1}{16}$(3)	$\frac{1}{16}$(4)	$\frac{1}{16}$(5)	$\frac{1}{16}$(6)
$\frac{1}{4}$	3		$\frac{1}{16}$(4)	$\frac{1}{16}$(5)	$\frac{1}{16}$(6)	$\frac{1}{16}$(7)
$\frac{1}{4}$	4		$\frac{1}{16}$(5)	$\frac{1}{16}$(6)	$\frac{1}{16}$(7)	$\frac{1}{16}$(8)

Table 14.4 Probability Distribution for $X_1 + X_2$

$x_1 + x_2$	$P(x_1 + x_2)$
2	$\frac{1}{16}$
3	$\frac{2}{16}$
4	$\frac{3}{16}$
5	$\frac{4}{16}$
6	$\frac{3}{16}$
7	$\frac{2}{16}$
8	$\frac{1}{16}$

appropriate diagonal. For example, $P(X_1 + X_2 = 4) = \frac{3}{16}$. Proceeding in this way, we obtain the probability distribution for $X_1 + X_2$ (see Table 14.4)

As a next step, consider the joint distribution of $X_1 + X_2$ and X_3 in a random sample of size 3, as shown in Table 14.5. The numbers in parentheses are values for the sum $X_1 + X_2 + X_3$ and, as before, the probability for any value of the sum is given by adding the probabilities on the appropriate diagonal. Thus,

$$P(X_1 + X_2 + X_3 = 7) = \frac{3}{64} + \frac{4}{64} + \frac{3}{64} + \frac{2}{64} = \frac{12}{64}$$

In this way we obtain the probability distribution for $X_1 + X_2 + X_3$ (see Table 14.6).

In an analogous manner we can obtain the probability distribution of $X_1 + X_2 + X_3 + X_4$, as shown in Tables 14.7 and 14.8.

Table 14.5 Joint Distribution of $X_1 + X_2$ and X_3

		$P(x_3)$	$\frac{1}{4}$	$\frac{1}{4}$	$\frac{1}{4}$	$\frac{1}{4}$
$P(x_1 + x_2)$	$x_1 + x_2$	x_3	1	2	3	4
$\frac{1}{16}$	2		$\frac{1}{64}$(3)	$\frac{1}{64}$(4)	$\frac{1}{64}$(5)	$\frac{1}{64}$(6)
$\frac{2}{16}$	3		$\frac{2}{64}$(4)	$\frac{2}{64}$(5)	$\frac{2}{64}$(6)	$\frac{2}{64}$(7)
$\frac{3}{16}$	4		$\frac{3}{64}$(5)	$\frac{3}{64}$(6)	$\frac{3}{64}$(7)	$\frac{3}{64}$(8)
$\frac{4}{16}$	5		$\frac{4}{64}$(6)	$\frac{4}{64}$(7)	$\frac{4}{64}$(8)	$\frac{4}{64}$(9)
$\frac{3}{16}$	6		$\frac{3}{64}$(7)	$\frac{3}{64}$(8)	$\frac{3}{64}$(9)	$\frac{3}{64}$(10)
$\frac{2}{16}$	7		$\frac{2}{64}$(8)	$\frac{2}{64}$(9)	$\frac{2}{64}$(10)	$\frac{2}{64}$(11)
$\frac{1}{16}$	8		$\frac{1}{64}$(9)	$\frac{1}{64}$(10)	$\frac{1}{64}$(11)	$\frac{1}{64}$(12)

Table 14.6 Probability Distribution for $X_1 + X_2 + X_3$

$x_1 + x_2 + x_3$	$P(x_1 + x_2 + x_3)$
3	$\frac{1}{64}$
4	$\frac{3}{64}$
5	$\frac{6}{64}$
6	$\frac{10}{64}$
7	$\frac{12}{64}$
8	$\frac{12}{64}$
9	$\frac{10}{64}$
10	$\frac{6}{64}$
11	$\frac{3}{64}$
12	$\frac{1}{64}$

The probability distributions for the sample total are plotted in Figure 14.2. As with the binomial distribution, it is startling that the bell-shaped phenomenon begins to occur with extremely small samples. With large samples from this original uniform distribution, the sum can, for all practical purposes, be regarded as normal.

Table 14.7 Joint Distribution of $X_1 + X_2 + X_3$ and X_4

		$P(x_4)$	$\frac{1}{4}$	$\frac{1}{4}$	$\frac{1}{4}$	$\frac{1}{4}$
$P(y)$	y^a	x_4	1	2	3	4
$\frac{1}{64}$	3		$\frac{1}{256}(4)$	$\frac{1}{256}(5)$	$\frac{1}{256}(6)$	$\frac{1}{256}(7)$
$\frac{3}{64}$	4		$\frac{3}{256}(5)$	$\frac{3}{256}(6)$	$\frac{3}{256}(7)$	$\frac{3}{256}(8)$
$\frac{6}{64}$	5		$\frac{6}{256}(6)$	$\frac{6}{256}(7)$	$\frac{6}{256}(8)$	$\frac{6}{256}(9)$
$\frac{10}{64}$	6		$\frac{10}{256}(7)$	$\frac{10}{256}(8)$	$\frac{10}{256}(9)$	$\frac{10}{256}(10)$
$\frac{12}{64}$	7		$\frac{12}{256}(8)$	$\frac{12}{256}(9)$	$\frac{12}{256}(10)$	$\frac{12}{256}(11)$
$\frac{12}{64}$	8		$\frac{12}{256}(9)$	$\frac{12}{256}(10)$	$\frac{12}{256}(11)$	$\frac{12}{256}(12)$
$\frac{10}{64}$	9		$\frac{10}{256}(10)$	$\frac{10}{256}(11)$	$\frac{10}{256}(12)$	$\frac{10}{256}(13)$
$\frac{6}{64}$	10		$\frac{6}{256}(11)$	$\frac{6}{256}(12)$	$\frac{6}{256}(13)$	$\frac{6}{256}(14)$
$\frac{3}{64}$	11		$\frac{3}{256}(12)$	$\frac{3}{256}(13)$	$\frac{3}{256}(14)$	$\frac{3}{256}(15)$
$\frac{1}{64}$	12		$\frac{1}{256}(13)$	$\frac{1}{256}(14)$	$\frac{1}{256}(15)$	$\frac{1}{256}(16)$

[a] $y = x_1 + x_2 + x_3$.

Table 14.8 Probability Distribution of $X_1 + X_2 + X_3 + X_4$

$x_1 + x_2 + x_3 + x_4$	$P(x_1 + x_2 + x_3 + x_4)$
4	$\frac{1}{256}$
5	$\frac{4}{256}$
6	$\frac{10}{256}$
7	$\frac{20}{256}$
8	$\frac{31}{256}$
9	$\frac{40}{256}$
10	$\frac{44}{256}$
11	$\frac{40}{256}$
12	$\frac{31}{256}$
13	$\frac{20}{256}$
14	$\frac{10}{256}$
15	$\frac{4}{256}$
16	$\frac{1}{256}$

14.7 We have illustrated with two simple examples the primitive idea of the central limit theorem. Actually, there are many different versions of the central limit theorem. Adams [1] summarizes the important versions and gives a fascinating account of the development of the theorem.

SUMMARY. A vast body of statistical theory and methods presupposes a **normal distribution.** Often the raw data are obviously not well described by a normal distribution, but analyses are performed on **sums** or **averages** of observations. These sums and averages tend to be normally distributed. A well-known limit theorem asserts this to be the case and, because of the importance to statistics, it is called the **central limit theorem.**

The **limiting normal** phenomenon can be observed by graphing the binomial distribution for increasing values of n or the Poisson distribution for increasing values of λ.

Not only does the distribution of a sum of random variables approach normality, but the **convergence** is also quite **rapid.** This is observed by obtaining the distribution of $X_1 + X_2, X_1 + X_2 + X_3, X_1 + X_2 + X_3 + X_4$, etc., when the observations are taken from a discrete, highly **nonnormal** distribution. Very quickly the distributions begin to acquire the symmetric bell shape.

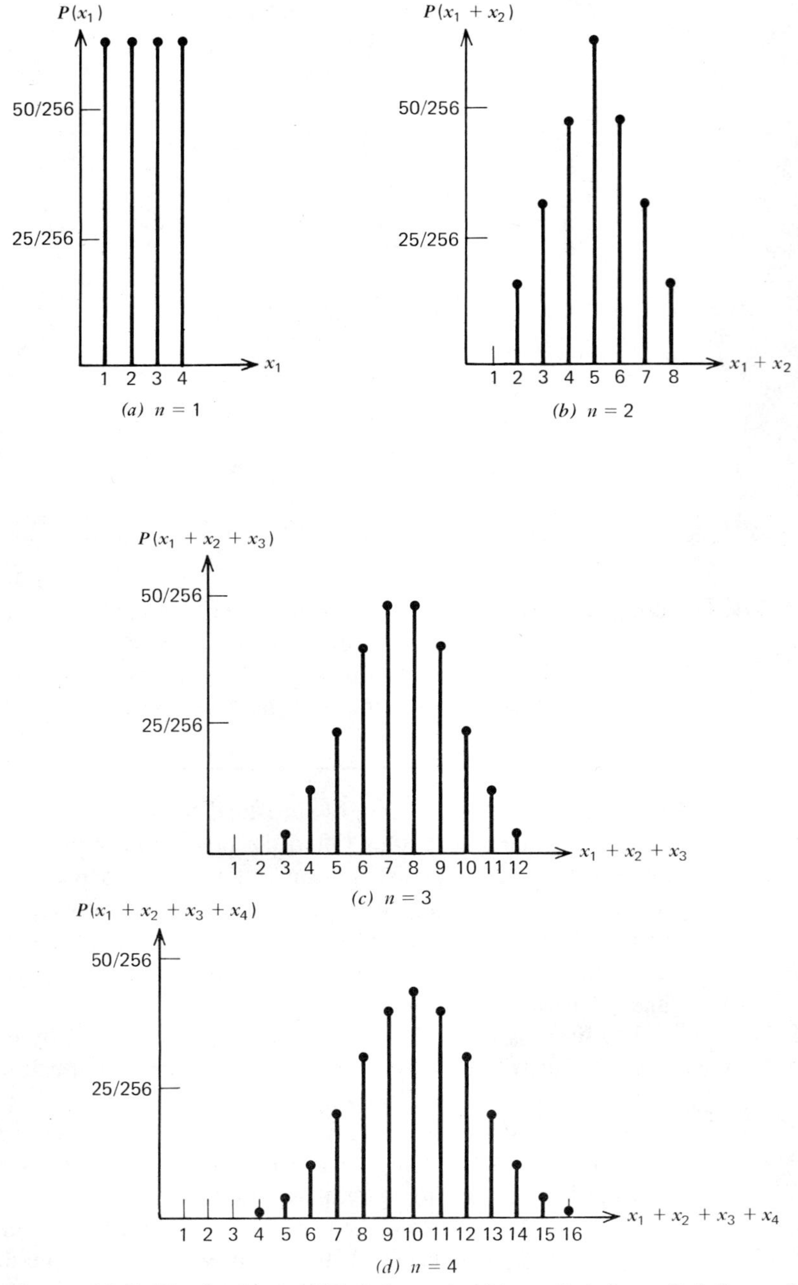

Figure 14.2 Distribution of ΣX. (*a*) $n = 1$. (*b*) $n = 2$. (*c*) $n = 3$. (*d*) $n = 4$.

REFERENCES

1. Adams, William J. (1974). *The Life and Times of the Central Limit Theorem*, New York: Kaedmon.
2. De Moivre, Abraham. (1933). *Approximatio ad Summam terminorum Binomii $\overline{a + b}^n$ Seriem expansi.*
3. Polya, G. (1920). *Uber den zentralen Grenzwertsatz der Wahrscheinlichkeitsrechnung und das Momentenproblem.*

EXERCISES

1. Extend Figure 14.1 by graphing the binomial probabilities for $p = \frac{1}{3}$, $n = 6, 7, 8, 9, 10$.
2. Given the following probability distribution:

x	$P(x)$
1	$\frac{4}{10}$
2	$\frac{3}{10}$
3	$\frac{2}{10}$
4	$\frac{1}{10}$

find the joint distribution of X_1 and X_2 and from that find the probability distribution of $X_1 + X_2$. Next find the distribution of $X_1 + X_2 + X_3$ and $X_1 + X_2 + X_3 + X_4$. Graph the distributions. Does the convergence to normality seem more or less rapid than with the uniform population of Section 14.6? Why is this reasonable?

3. Find the probability distribution for the sum of a random sample of size 5, starting with the population distribution given in Table 14.2.
4. A student asked to discuss the central limit theorem on an exam wrote, "As the number of samples increases, $\bar{x}$ becomes more nearly normal." What is wrong with this answer?
5. A sample of size 10 is taken from a large population so that $\sigma_{\bar{x}}^2 = \sigma^2/n$. If $\mu = 8$ and $\sigma^2 = 40$, standardize the sample mean value of $\bar{x} = 12$ by subtracting μ and dividing by the standard deviation of $\bar{x}$. If μ is approximately normal, use the standardized value and Table A.3 to find the probability that $\bar{x} < 12$.

6. If the mean of a sample of size 16 from a population with $\mu = 15$ and $\sigma^2 = 25$ is approximately normal, find the following probabilities.
   a. $P(\bar{x} \geq 14)$.
   b. $P(13 \leq \bar{x} \leq 16)$.
   c. $P(\bar{x} \leq 14.5)$.
7. A busy intersection is observed over a long period of time, and the number of arrivals at a red light is found to be a Poisson variable with parameter $\lambda = 5.2$. We are interested in the average number of arrivals ($\bar{x}$) for 36 ($n = 36$) red lights. Use the central limit theorem and Table A.3 to find the probability that $\bar{x}$ exceeds 6.5.
8. A five-point rating scale is used on a questionnaire. If all of the responses -2, -1, 0, 1, and 2 are equally likely, find the mean response μ and the variance σ^2 of the population of responses. What would be the mean and variance of the average response, $\bar{x}$, on a questionnaire with 50 questions? Find the probability that $\bar{x} > 1$.
9. Instead of the five-point scale used in Exercise 8, use the five-point scale 1, 2, 3, 4, 5. What is the mean and variance if all scores are equally likely? What is the mean and variance of the average response, $\bar{x}$, on a questionnaire with 50 questions? Find the probability that $\bar{x} > 3$.
10. As n becomes larger, the binomial probabilities are more closely approximated by normal distribution probabilities. For $n = 100$ and $p = \frac{1}{2}$, find the probability that $X = 50$. If the histogram with classes -0.5–0.5, 0.5–1.5, etc., is used, this probability is given by the area of the bar of unit width centered on the class mark 50. Approximate the binomial probability by the area under the normal curve from 49.5 to 50.5 with $\mu = 50(np)$ and variance $25(npq)$.
11. For large n the binomial distribution is well approximated by the Poisson and normal distributions. Therefore the Poisson distribution with large λ should be well approximated by the normal distribution. Graph the Poisson probability distribution for $\lambda = 10$, $20, \ldots, 50$.
12. For $\lambda = 20$, represent the Poisson distribution with a histogram using class marks $1, 2, 3, \ldots$. Approximate the area of the bar with class mark 20 by the area under the normal curve from 19.5 to 20.5 using $\mu = \lambda = 20$ and $\sigma^2 = \lambda = 20$.
13. Draw 150 random samples of five one-digit numbers from Table A.1. For each sample calculate $\bar{x}$. Construct the frequency distribution and construct a histogram for your samples. Note that the distribution of $\bar{x}$ is much more normal than the distribution of individual xs.

14. An instrument is used in an educational research experiment that consists of binary responses (yes-no, true-false, etc.) to 100 items. One of the responses is labeled zero and the other one. In a large population of equally likely zeros and ones, what is the mean and variance? If the responses on each item are equally likely, what is the mean and variance of the average response on 100 items? Using the central limit theorem, what is the probability that $\bar{x} > .75$?
15. The graduate committee of a statistics department claims that its new students are definitely better than average. The national mean and standard deviation on the Graduate Record Examination analytical score are known to be 583 and 117, respectively. The 20 new students admitted to the department have an average analytical score of 624. If the 20 students were a random sample from the population, what is the probability of an average score as great as 624?
16. After several dry years in a certain city, people are predicting a major shift in weather patterns. The annual rainfall for the past 50 years has a mean of 42.36 inches and a standard deviation of 6.11 inches. The average rainfall for the last 5 years has been 35.72 inches. What is the probability of a sample mean this small using a sample of size 5 from a population with mean 42.36 and standard deviation 35.72? Does this probability cast doubt on the hypothesis of a major weather change?

ST. JAMES'S GATE BREWERY. W. S. Gosset went to work here in 1899. Arthur Guinness Son and Company, Ltd.

15

"STUDENT'S" t

15.1 Arthur Guinness set up a business as a brewer in 1859 on property near St. James's Gate, an ancient entrance to the city of Dublin [3]. The firm prospered, becoming a limited liability company in 1886. The Guinness family became enormously wealthy and was a great public benefactor. By the late 1800s, the Guinness firm, like many industries today, had begun to hire a number of university-trained scientists.

W. S. Gosset was born in Canterbury, the ancient cathedral city, in 1876. After his education at Winchester and at New College, Oxford, he joined the firm of Arthur Guinness Son and Company, Ltd., in Dublin as a brewer in 1899. Soon after his employment by Guinness, he began to apply his creative genius to interpretation of data and the planning of barley experiments. After several years in Dublin, Gosset was sent to London for a year's specialized study. He spent 1906 to 1907 at University College London, where he studied statistics with Karl Pearson.

15.2 In 1907 Gosset published his first paper under the pen name or pseudonym "Student." We mentioned this paper in our earlier lecture on the Poisson distribution. In 1908 there appeared in *Biometrika* the second paper by Gosset [6] entitled "The Probable Error of a Mean." It contains several important results, including the distribution of a statistic that has become known as **"Student's"** t.

Gosset published a series of important statistical papers, always under the pen name "Student." Initially his true identity was known to only a few people; most readers of his early papers could only speculate about his identity. E. S. Pearson [4] notes that for many years after the 1907 paper, an air of romance was associated with the name "Student." Neyman [5] indicates that even after the death of Gosset in 1937, many statisticians did not know that he was "Student."

15.3 What was Gosset's contribution? In trying to answer this question, it is important to remind ourselves of some of the well-known results at that time. It was already known that if we sample from a normal population with mean μ and variance σ^2, the sampling distribution of $\bar{x}$ is normal with mean μ and variance σ^2/n. It was also known that $\sqrt{n}(\bar{x} - \mu)/\sigma$ is a **standard normal variable**, so that the standard normal table is useful in assessing the meaning of a sample average $\bar{x}$. For example, suppose that a variety of barley in common use in a region gives a yield with a standard deviation of 6 (in bushels per acre). A new variety of barley is to be used on 10 test plots, and we want some way of assessing the results to help us decide whether the new variety is different or not. A usual procedure at that time was to determine the interval within which a variable would fall with probability $\frac{1}{2}$. Now, from Table A.3, we can see that

$$P(-.6745 \leq Z \leq .6745) = .5$$

so that

$$P\left(-.6745 \leq \frac{\bar{x} - \mu}{\sigma/\sqrt{n}} \leq .6745\right) = .5$$

Since $\sigma = 6$ and $n = 10$, we have that

$$P\left(-.6745 \leq \frac{\bar{x} - \mu}{6/\sqrt{10}} \leq .6745\right) = .5$$

By multiplying all terms of the inequality by $6/\sqrt{10}$, we obtain

$$P(-1.28 \leq \bar{x} - \mu \leq 1.28) = .5$$

In other words, the probability that the sample mean will differ from the population mean by more than 1.28 is .5. If the experiment produced a sample average of 30 bushels per acre, we would feel the odds are even that μ lies in the interval given by 30 ± 1.28.

15.4 Gosset's work dealt with **small samples**, frequently in situations where it was unreasonable to assume that σ was known. He was led to consider the distribution of

$$Z = (\bar{x} - \mu)/s^*$$

where

$$s^{*2} = \sum_{i=1}^{n} (x_i - \bar{x})^2/n$$

and where the parent population is normal with mean μ and variance σ^2 (Z in this case is not the standard normal). Although he did not complete the mathematical proof, he did correctly infer that s^{*2} is a constant times a **chi-square variable** and that $\bar{x}$ and s^{*2} are **independent**. From these results he reached the correct conclusion that the sampling distribution of Z is given by

$$f(z) = c(1 + z^2)^{-n/2}$$

where c is a positive constant. By drawing many samples and calculating Z for each sample he was able to demonstrate good agreement between the theoretical curve and the sampling data. In today's terminology we would say that Gosset demonstrated the form of the sampling distribution by simulation.

15.5 Eisenhart [1] traces the gradual transition from z to t where

$$t = \sqrt{n}(\bar{x} - \mu)/s \quad \text{and} \quad s^2 = \sum_{i=1}^{n} (x_i - x)^2/(n - 1)$$

where the parent population is normal with mean μ and variance σ^2. Then the sampling distribution of t is given by

$$f(t) = c\left(1 + \frac{t^2}{n - 1}\right)^{-n/2}$$

Thus the distribution of t depends on n, whereas the distribution of $\sqrt{n}(\bar{x} - \mu)/\sigma$ does not. It has become customary to describe this dependence in terms of $n - 1$, called **degrees of freedom**, instead of in terms of n. Figure 15.1 shows the t curve for several values of $n - 1$ as well as a standard normal curve. From these curves we can see that (1) the t curves are symmetric bell-shaped curves, (2) the t curves have greater variability than the standard normal curve, and (3) the t curves appear to approach the standard normal curve as $n - 1$ (equivalently n) approaches infinity. This limiting property can be shown mathematically. The t distribution is tabled in Table A.5. To illustrate the table,

$$P(t < 2.228) = .975$$

for 10 degrees of freedom.

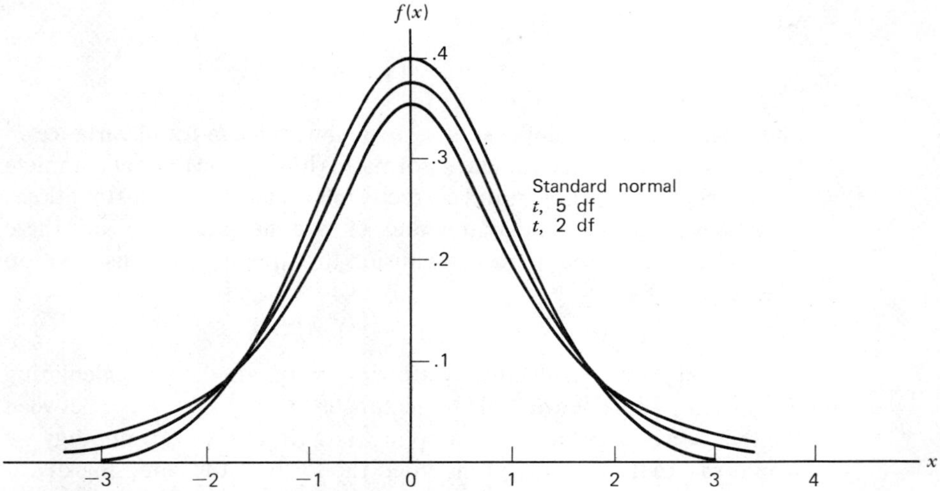

Figure 15.1 Family of "Student's" t Curves.

15.6 To understand the impact "Student's" t distribution made on small sample statistics, consider the barley yield experiment mentioned earlier and suppose that we had no idea of a reasonable value for σ. For a sample size of $n = 10$ (9 degrees of freedom), we know that

$$P(-.703 \leq t \leq .703) = .5$$

$$= P\left(-.703 \leq \frac{\bar{x} - \mu}{s/\sqrt{n}} \leq .703\right)$$

$$= P(-.703s/\sqrt{n} \leq \bar{x} - \mu \leq .703s/\sqrt{n})$$

$$= P(|\bar{x} - \mu| \leq .703s/\sqrt{n})$$

Then we feel the chances are even that $\bar{x}$ will differ from μ by more than $.703s/\sqrt{n}$. If we observe $\bar{x} = 35$, $n = 10$, $s^2 = 10$ so that

$$.703s/\sqrt{n} = .703$$

we would doubt that μ could be 30, for example.

15.7 It is probably difficult to overestimate the importance of Gosset's 1908 paper. Not only is the t distribution important in its own right, but it ushered in a whole new era of small sample statistics.

SUMMARY. W. S. Gosset was hired by Guinness breweries in 1899. He was led to his work in statistics by the requirements of his job. His paper, published in 1908 under the pseudonym **"Student,"** was of fundamental importance to science and profoundly affected the development of statistics.

Probability statements about the sample mean of normal observations were already possible **if μ and σ were known**. This allowed assessment of plausible values for μ once $\bar{X}$ was observed. If the **sample size** were **large**, the **sample variance** could be used in place of σ^2, but no one dared try **small samples**.

Gosset replaced σ^2 by $s^{*2} = \Sigma(x - \bar{x})^2/n$ and proceeded to find the distribution of $Z = (\bar{x} - \mu)/s^*$. He inferred the correct distribution, almost completed its derivation, and verified it by **simulation**.

As the result of correspondence with Fisher, Gosset's result gradually evolved to its present form with $s^2 = \Sigma(x - \bar{x})^2/(n - 1)$ instead of s^{*2} and $t = \sqrt{n}(\bar{x} - \mu)/s$.

REFERENCES

1. Eisenhart, Churchill. (1979). "On the Transition from 'Student's' *z* to 'Student's' *t*," *The American Statistician, 33*, 6–10.
2. Fisher, R. A. (1939). "Student," *Annals of Eugenics*, 9, 1–9.
3. Guinness Co. (1939). *Guinness*, Leeds: John Waddington, Ltd.
4. McMullen, Launce, and E. S. Pearson. (1939). "William Sealy Gosset, 1876–1937," *Biometrika, 30*, 205–250.
5. Neyman, J. (1938). "Mr. W. S. Gosset," *Journal of the American Statistical Association, 33*, 226–228.
6. "Student." (1908). "The Probable Error of a Mean, *Biometrika*, 6, 1–25.

EXERCISES

1. Using Table A.5, find:
   a. $P(t > 2.228)$ with 10 degrees of freedom.
   b. $P(t > 2.086)$ with 20 degrees of freedom.
   c. $P(-1.341 < t < 2.131)$ with 15 degrees of freedom.

d. t_0 where $P(t < t_0) = .05$ with 20 degrees of freedom.
e. t_0 where $P(|t| > t_0) = .05$ with 15 degrees of freedom.
f. $P(|t| > 2.179)$ with 12 degrees of freedom.

2. Using Table A.5 and the fact that the distribution of t is symmetric about zero, graph the cumulative t distribution for degrees of freedom 5, 10, 15, 20, 30, and 120. On the same graph, show the cumulative standard normal distribution. Note that the t curves approach the standard normal curves as the degrees of freedom become large.
3. Given the set of data 8, 10, 12, 10, 8, 9, 8, 12, 13, 12:
   a. Calculate $\Sigma(x - \bar{x})^2$ by (1) calculating $\bar{x}$, (2) subtracting $\bar{x}$ from each x to get $x - \bar{x}$, (3) squaring each $x - \bar{x}$, and (4) summing.
   b. Calculate $\Sigma(x - \bar{x})^2$ by (1) squaring each x and then adding to get Σx^2, and (2) then subtracting $(\Sigma x)^2/n$.
   c. Calculate $\bar{x}$ and s^2.
   d. Use "Student's" t to calculate $P(|\bar{x} - \mu| > 2s/\sqrt{n})$.
4. Given the random sample from a normal population 22.3, 17.1, 18.5, 19.6, 22.5, 21.0, 20.9:
   a. Calculate $\bar{x}$ and s^2.
   b. Use "Student's" t to calculate $P(|\bar{x} - \mu| > 2s/\sqrt{n})$.
5. The use of t tables is sometimes difficult because of notation. Often t_α denotes a t value such that $P(|t| > t_\alpha) = \alpha$. Thus, if the degrees of freedom are 10, we have from Table A.5 that $P(|t| > 2.228) = .05$. So $t_{.05} = 2.228$ with 10 degrees of freedom. What is the subscript on t for the following?
   a. 0.691 with 15 degrees of freedom.
   b. 2.624 with 14 degrees of freedom.
   c. 4.144 with 10 degrees of freedom.
   d. 2.571 with 5 degrees of freedom.
6. Twelve students chosen at random from the undergraduate students have the following weights (pounds): 141, 196, 167, 174, 172, 159, 148, 157, 161, 178, 164, and 155. Assuming that the population mean is 165, calculate the t value for this sample. Convert the weights to kilograms and recalculate the t value. It should be the same.
7. Verify algebraically that if every observation x is replaced by $y = ax + b$ and μ is replaced by $a\mu + b$, the t value remains the same.
8. Members of a horse club have tried various medical and grooming aids in preparing their horses for show competition. Various products are available to stimulate hoof growth; the club decides to study the effect on hoof growth of one particular product. Each member of the club uses the product on two hooves of his or her horse. The growth is

measured weekly and, at the end of the trial period, the increase in growth over the untreated hooves is recorded for each horse. The growth (in inches) for the 14 horses is as follows: 1.1, -1.0, 2.2, 1.4, 0.6, 0.1, 0.3, 0.4, 1.0, 0.7, -0.6, -0.2, 0.4, and 0.5. Calculate the t value for a μ of zero. Find the probability of t being this large in absolute magnitude. Do you feel that the data support the claim that the product stimulates hoof growth?

OPINIONS, CONCLUSIONS, AND DECISIONS

W. S. GOSSET. 1876–1936. The Granger Collection.

16

DEDUCTIONS AND INFERENCES

16.1 *"Interesting, though elementary," said he as he returned to his favourite corner on the settee. "There are certainly one or two indications upon the stick. It gives us the basis for several deductions."**

The Hound of the Baskervilles

In case after case, as described by the faithful Watson, the legendary Sherlock Holmes exercised the principles of **logic**, always arriving at the correct conclusion. The unfortunate Watson (and the reader) had less success. Holmes reigned supreme, the personification of the man of logic. In reality, the feats of logic of actual human beings have been far more dramatic. The stories of the countless legion of men and women of science are more incredible and more romantic than the tales of Holmes and Watson.

In the quest for knowledge the workers of science, using the rules of logic, constantly try to reach conclusions from the data available to them. But what are the rules of logic?

16.2 Logic is the science that attempts to systematize the process of judging that a conclusion is justified given that certain other statements are true. Most propositions that we accept as true or probable cannot be so judged on the basis of direct sensory perception. For example, our acceptance of numbers such as the speed of light, the speed of sound, the distance to the moon, etc., does not rest on direct observation for most of us. Instead, our acceptance of these numbers rests on acceptance of a

* Published with the permission of the copyright owner of the Sir Arthur Conan Doyle literary estate.

host of other propositions, many not explicitly stated or perceived. In fact, most propositions are justified by an argument that proceeds from a given set of statements called **premises** to a concluding statement called a **conclusion**.

16.3 The process of reaching a conclusion from stated premises is often called **inference** and, for the sake of simplicity, the study of inference usually denotes two types: **deductive** inference (deduction) and **inductive** inference (induction). The word inference is also often used to denote inductive inference only. Following Skyrms [1], we will say that a statement is a factual claim, an argument is a list of statements beginning with premises and ending with a conclusion, and logic is the study of the evidential link between the premises and the conclusion.

To illustrate the idea of a deductive argument, consider the following.

Example 1: Deductive Argument.
Premises: John's birthday is in the month of September.
It rained on John's last birthday.
Conclusion: It rained in September.

If we accept the premises in Example 1 as being true, we must accept the conclusion as being true. There is no escape from this. The conclusion follows with certainty from the premises. This argument illustrates the idea of a deductively **valid** argument. In general, an argument is said to be valid if, when the premises are true, the conclusion must be true. It should be noted that validity is not a measure of the certainty with which the conclusion follows from the premises.

Example 2: Deductive Argument.
Premises: John's birthday is in the month of Octaroon.
It rained on John's last birthday.
Conclusion: It rained in Octaroon.

The argument in Example 2 is deductively valid. Is it true? No one knows. But, given that the premises are true, the conclusion follows with absolute certainty. Let us now consider an argument that is not deductively valid.

Example 3: Inductive Argument.
Premises: Last September was the rainiest on record.
John's birthday is in September.
Conclusion: It rained on John's last birthday.

The conclusion in Example 3 does not follow with certainty given that the premises are true and is not, therefore, deductively valid. It is, however, quite **plausible** and, in some parts of the country, would be quite likely. In any locale where the rainfall is very low it would be quite unlikely. This example does serve as an illustration of an inductive argument. A basic problem in inductive inference is to devise ways of measuring the strength of an inductive argument.

Most approaches to measuring the strength of an inductive argument have used probability in some way. Readers interested in such approaches should be reminded at this point of the two divergent interpretations of probability: probability as a **measure of belief** and probability as a **physical constant** in some system. To those embracing the first interpretation, it has seemed only natural to define the strength of an inductive argument as

$$P(\text{conclusion true} \mid \text{premises true})$$

To those embracing the second interpretation of probability, it has seemed just as natural to reject this but to devise other measures that involve probability. In statistics, these two approaches have led to **Bayesian** and **non-Bayesian** statistics, respectively.

16.4 It should be quite clear that a large part of statistics is concerned with the process, called **statistical inference**, of drawing conclusions about the population on the basis of information obtained from a sample from the same population. It should also be clear at this point that such a process is a process in inductive inference. From this point of view statistical inference lies in the center of all experimental science. In anticipation of lectures to follow, we now want to give an overview of some of the main topics of statistical inference.

16.5 Many problems in statistical inference are motivated by real problems in science. The **statistical methods** used are based on research that has assumed that the **sample** has come from a **population** well described by some particular model. For example, suppose we are interested in the relative effectiveness of two weight-reducing diets and we have data on weight loss from a sample of people for each of the diets. We **may** assume that the samples of weight losses are random samples from two populations of weight losses of people and that the two populations are well described by two normal populations with means μ_1 and μ_2, respectively. Our methods are then based on a mathematical **model** of a random sample from each of two normal populations. It should be apparent that the choice of a model is a very important step. What is the process by which we choose a model or conclude that a model is highly plausible?

The **choice of model** is itself an exercise in inductive inference and, if the choice is based on data, it is an exercise in statistical inference. An array of statistical methods has been devised to assist in the choice of a model; they range from simple graphical procedures to formal procedures called **goodness-of-fit tests**. Goodness-of-fit tests are so named because they give the user help in judging how well a population is described (or fitted, like a suit of clothes) by a given model.

16.6 Once a model has been accepted as reasonable, it is natural to focus attention on the unknown constants, called parameters, that appear in the model. With a normal distribution we are concerned with μ and σ^2, with a Poisson distribution with λ, etc. Formal procedures (called tests) have been devised for judging how well hypothesized values of the parameters are supported by the sample data. The tests are usually called **tests of significance** and (or) **tests of hypotheses**.

16.7 Quite often, particularly when the parameters of the mathematical model have a physical interpretation, it will be necessary to specify or guess a value for a parameter on the basis of the sample data. For example, if μ of a normal distribution represents the average family income for a community, it may be necessary to make a guess for the value of μ, based on a random sample of family incomes. Guesses of parameter values that are based on sample data are called **estimates**. Estimates should be obtained from procedures that have known properties and that have been tried and proven in applications.

Obviously, any estimated (or guessed) value for a parameter has an uncertainty attached to it, and one way of describing such uncertainty is by giving an **interval** of values for a parameter. Various methods of obtaining intervals have been studied; the best-known intervals are **confidence intervals** and **Bayesian intervals**.

16.8 We have given an overview of some of the main branches of statistical inference here. We have indicated that statistical inference lies at the heart of experimental science. It is worth noting that statistical inference rests on the assumption of a mathematical model. Without knowledge of such models, statistical inference is a fairly narrow subject.

SUMMARY. Logic is the study of the link between statements that strengthens the justification of some when others are true. The process of reaching a **conclusion** from a given set of **statements** called premises is often called **inference**. Inference falls into two broad classes: **deductive** and **inductive**. Deductive inference is sometimes described as reasoning from the general to the particular, while inductive inference is from the particular to the general. With deductive inference the concern is with the **validity** of the argument. With inductive inference the concern is with the **plausibility** of the conclusion given that the premises are true.

Statistical inference is concerned with drawing conclusions about the population on the basis of a sample. One of the first problems is the choice of a mathematical **model** for the random sample. Usually the next problem is the choice of parameter values in the model chosen. **Bayesian** statistical methods are directed toward finding probability statements about the parameters. **Non-Bayesian** methods, using a different interpretation of probability, are concerned with **estimates** of the parameters, **tests** for hypothesized parameter values, and **intervals** of reasonable parameter values.

REFERENCE

1. Skyrms, Brian. (1966). *Choice and Chance*, Belmont, Calif.: Dickenson.

EXERCISES

1. In your daily newspaper note examples of arguments that are deductively valid and of arguments that are inductively strong.
2. Note important arguments from your major field that are not deductively valid but that are inductively strong.
3. How would you classify the argument that concludes that $\mu \neq 10$ given the premises that $\bar{x} = 15$, $\sigma = 6$, and $n = 36$? How would you measure the strength of the argument?
4. How would you classify the argument that μ lies between 25 and 30 given the premises that $\bar{x} = 27.5$, $\sigma = 4$, and $n = 20$, and the sample is from a normal distribution? How would you measure the strength of the argument?

5. A company has discriminatory practices against women as a group in hiring and promotion of employees. Theo is a woman who formerly worked for the company. The company has discriminated against Theo. Is the previous argument a deductively valid argument? Is it inductively plausible?
6. A spacecraft is programmed to return to earth intact. The chance of its striking human beings is calculated to be less than 1 in 1 million. Therefore it is quite unlikely that it will land in any metropolitan area. Is this a deductively valid argument? Is it inductively plausible?
7. In June, before announcement of candidacy, a particular candidate is preferred by 10% of the people polled. In November, after the announcement, 12% of the people polled prefer the candidate, who claims to have gained support. Is the claim justified? How would you measure the strength of the claim?
8. A coffee manufacturer claims that its coffee yields more cups to the pound than a competitive brand, which advertises 250 cups per pound. Ten pounds are purchased by the coffee club of a department; after using the first pound to determine the amount required for the proper taste, the number of cups per pound is recorded for the last 9 pounds, giving 248, 254, 256, 253, 248, 250, 252, 246, and 256. Calculate the t value using a μ value of 250. What is the probability of a t this large or larger? Do the data support the idea that the new brand gives more than 250 cups per pound?

JERZY NEYMAN. University of California, Public Information.

17

STATISTICAL TESTS

17.1

You will see something new.
Two things. And I call them
*Thing One and Thing Two.**

The Cat in the Hat by Dr. Seuss

The first occurrence of a **statistical test** is uncertain, but an early example is given by Arbuthnot [1]. He considered the period of 82 years for which records were available on the number of males and females born in London and noted that in each year males exceeded females. He noted that under a "chance" **hypothesis**, the probability of such an outcome would be $(\frac{1}{2})^{82}$.

A second example of a statistical test is given by Stigler [5] who describes the trial of the Pyx, a test that has been used by the Royal Mint in London for about 800 years. During these years, coins have been taken from those minted and placed in a box called the Pyx. At specified times the box would be opened and the contents counted, weighed, and assayed for the official purpose of determining that the coinage issued by the Royal Mint met specifications. The antiquity of the trial of the Pyx is well documented, and Stigler offers a good bibliography. Before the twentieth century certain statistical tests were known and used (e.g., the chi-square test of Pearson [4]), but only in the present century were statistical tests fully developed. The early papers of "Student" gave an impetus to the development of a theory of tests, but Fisher, in a series of papers from 1915 onward, gave statistical testing a great push forward. The next major thrust began with the publication in 1933 of the important paper by Neyman and Pearson [3].

17.2 As conceived by Fisher [2, p. 13], a **test of significance** consisted of the following items.

1. A **hypothesized value** for a parameter of an assumed probability distribution.
2. A **test statistic** whose distribution is completely known when the hypothesized value is the true parameter value.
3. A **random sample** from the population.
4. **Calculation** of the statistic.
5. Calculation of the **probability** of the test statistic being as extreme as actually observed when the hypothesis is true.

The probability calculated in item 5 is called the **significance level** or **observed significance level**. If it is small we have the logical disjunction: either the hypothesis is false or the hypothesis is true and an event with low probability has occurred.

Example 1. A nonprescription remedy for cough and nasal congestion indicates on the label that it is 5% alcohol. A consumer believes that the amount of alcohol varies from bottle to bottle but that the average amount in all bottles produced may be in the neighborhood of 5%. He buys 10 bottles and performs a laboratory analysis on each bottle to find the amount of alcohol. The results in percent are 5.01, 4.87, 5.11, 5.21, 5.03, 4.96, 4.78, 4.98, 4.88, and 5.06. If it is believed that the amount of alcohol is normally distributed, do the data support the hypothesized mean value of 5%?

1. **Hypothesis.** $\mu = 5$
2. **Test Statistic.**

$$t = \frac{\bar{x} - 5}{s/\sqrt{10}}$$

   Note that if $\mu = 5$, the distribution of this test statistic is completely known and tabulated ("Student's" t with 9 degrees of freedom).
3. **Random Sample.** Ten numbers given by the laboratory analyses.
4. **Calculation of Test Statistic.**

$$\Sigma x = 49.89$$

$$\bar{x} = 4.989$$

$$\Sigma x^2 = 249.0425$$

$$\Sigma(x - \bar{x})^2 = 0.14129$$

$$s^2 = 0.0156989$$

$$s_{\bar{x}}^2 = 0.00156989$$

$$\bar{s}_{\bar{x}} = 0.0396$$

$$t = \frac{4.989 - 5}{0.0396}$$

$$= -0.2778$$

5. **Calculation of Observed Significance Level.** From Table A.5 we see that the probability of a t value this extreme (as large as $+0.2778$ or as small as -0.2778) is about .80. This probability is very large and suggests that a mean value of 5% is quite consistent with the data. If, on the other hand, we had obtained a very small probability, we would have been forced to conclude that either the 5% mean was false or that an event with low probability had occurred.

Note that the observed significance level does **not** give the probability that the null hypothesis is true. It is **not** a probability statement of any kind about the parameter (or parameters). Consider the following analogy. An accused person on trial pleads "guilty," and experience indicates that only 10% of innocent prisoners plead guilty. Then, given the guilty plea, we feel that the prisoner is either guilty or innocent and an event with probability $\frac{1}{10}$ has occurred. However, this is not the same as saying that the probability of innocence is $\frac{9}{10}$. To state the matter in symbols,

$$P(\text{"not guilty" plea} \mid \text{innocent}) = .90$$

$$\neq P(\text{innocent})$$

The observed significance level is reported in many research journals and is often labeled simply the ***P* value**. In addition, many packaged statistical programs calculate the observed significance level routinely.

17.3

In 1933 Neyman and Pearson published a paper on statistical tests [3]. In subsequent years they and their followers developed the theory of tests in a manner similar to but more formal than that of Fisher. As developed by Neyman and Pearson, a statistical test is a decision procedure for accepting one of two possible hypotheses about a parameter. One of the hypotheses is called a **null hypothesis** and the other an **alternative hypothesis**. If one then uses a procedure to **accept** or **reject** the null hypothesis, there are four possible contingencies, as indicated in Table 17.1.

Table 17.1 Possibilities in Testing

	Decision	
State of Nature	Accept Null Hypothesis	Reject Null Hypothesis
Null hypothesis is true	No error	Type I error
Null hypothesis is false	Type II error	No error

If we reject the null hypothesis when it is false or accept it when it is true, no error in judgment occurs; otherwise, we commit one of two possible errors. These are called **Type I error** and **Type II error**. The error of rejecting the null hypothesis when it is true is a Type I error, and the error of accepting the null hypothesis when it is false is a Type II error.

Naturally, we would like to eliminate both errors, but this is impossible. Instead, we are forced to settle for controlling the probability of such errors. We now introduce a little more terminology of Neyman-Pearson hypothesis testing.

α = probability of Type I error

= significance level

β = probability of Type II error

$1 - \beta$ = **power** of the test

We would like both α and β to be small, but they seem to have minds of their own. When we make α smaller, β tends to become larger; when we make β smaller, α tends to become larger. The simple procedure usually adopted is to fix α arbitrarily and to make β acceptably small by increasing the sample size.

Using the trial-by-jury analogy, we might take the null hypothesis to be "not guilty." Under Anglo-Saxon law and traditions, the probability of a Type I error is kept very small, with the consequence that the probability of a Type II error is sometimes larger than would be desired.

The components of a test following the Neyman-Pearson procedure are as follows.

1. **Null** and **alternative** hypotheses for a parameter of an assumed probability distribution.
2. α and β or α and n.
3. A **test statistic**.
4. A **decision rule**.
5. A **random sample** from the population.
6. **Calculation** of the test statistic.
7. A **decision** either to accept the null hypothesis or to reject the null hypothesis obtained from the calculated value of the test statistic and the decision rule.

Example 2. A manufacturer of a type of cracker indicates on each box that the box contains 269 grams. He is anxious not to ship boxes that have less weight than this and the quality control department designs a test procedure with the following components, assuming that the weight is a normal variable:

1. **Null Hypothesis.** $\mu = 269$.
   Alternative Hypothesis. $\mu < 269$.
2. $\alpha = 0.05$
   n is set at $n = 30$ after some consultation.
3. **Test Statistic.**

$$t = \frac{\bar{x} - 269}{s_{\bar{x}}}$$

4. **Decision Rule.** Reject the null hypothesis if $t < -1.699$.
5. **Random Sample.** Thirty boxes are selected at random from the production line, resulting in a mean $\bar{x} = 268$ and $s = 1.8$.
6. **Calculation of the Test Statistic.**

$$s_{\bar{x}} = \frac{s}{\sqrt{n}} = \frac{1.8}{\sqrt{30}} = 0.3286$$

$$t = \frac{\bar{x} - 269}{s_{\bar{x}}} = \frac{-1}{0.3268}$$

$$= -3.06$$

7. Because the calculated t is less than -1.699, we reject the null hypothesis and conclude that the production line is not meeting the advertised claim of 269 grams per box on the average.

17.4 In many cases a statistical test is used more or less to assess the data and not to reach any firm decision. This seems to have been the attitude of Fisher, who was thinking of research situations. In other cases the background of the problem demands a clear decision. This sort of situation seems to be better handled with some sort of decision rule, such as the Neyman-Pearson approach. We will discuss the matter further in Lecture 33.

SUMMARY. Traces of **statistical tests** are very old. One of the earliest is the **trial of the Pyx** in London, a test applied to coins by the Royal Mint for over 800 years. However, in the twentieth century, statistical tests have been studied, refined, and used widely. Theoretical development began with the work of **Pearson** in 1900 and the extremely important work of **"Student"** in 1908. **Fisher** contributed heavily to the development of tests from 1915 on.

Given a **hypothesized** value for a parameter, a **test statistic** is calculated from a **random sample** of observations. The **observed significance level** is the probability that a statistic as extreme as the one observed could have occurred by chance when the hypothesis is true. This value is a measure of plausibility of the hypothesized value.

Neyman and Pearson formulated a test as a more formal accept-reject rule. This theory requires specification of both a **null** and **alternative** hypothesis. Rejecting the null hypothesis when true and accepting the alternative hypothesis when false are called **Type I** and **Type II** errors, respectively. Although these errors cannot be avoided, tests can be designed to control their probabilities.

REFERENCES

1. Arbuthnot, J. (1710). "An Argument for Divine Providence Taken from the Constant Regularity of the Births of Both Sexes," *Philosophical Transactions of the Royal Society, 27*, 186–190.

2. Fisher, R. A. (1956). *Statistical Methods and Scientific Inference*, Edinburgh: Oliver & Boyd.

3. Neyman, J., and E. S. Pearson. (1933). "On the Problem of the Most Efficient Tests of Statistical Hypotheses," *Philosophical Transactions of the Royal Society, A-231*, 289–337.

4. Pearson, K. (1900). "On a Criterion that a Given Set of Deviations from the Probable in the Case of a Correlated System of Variables Is Such That It Can Be Reasonably Supposed to Have Arisen in Random Sampling," *Philosophical Magazine*, *50*, 157–176.
5. Stigler, Stephen M. (1977). "Eight Centuries of Sampling Inspection: The Trial of the Pyx," *Journal of the American Statistical Association*, *72*, 493–500.

EXERCISES

1. Given a random sample of size 25 from a normal population with mean $\mu = 10$ and $\sigma = 4$, find

$$P\left(\frac{\bar{x} - \mu}{\sigma/\sqrt{n}} > 2\right)$$

2. If $\bar{x} = 12$ for the sample of Exercise 1, find

$$Z_{\text{observed}} = \frac{12 - 10}{4/\sqrt{25}}$$

   What is $P(Z > Z_{\text{observed}})$?

3. We are given a random sample of size 25 from a normal population with unknown mean, μ, and $\sigma = 4$. It is hypothesized that the mean is 10, but there is some concern that it may be greater than 10. Calculate the observed significance level if $\bar{x} = 12$. Do the data support the hypothesized value of μ?
4. We have designed a sampling inspection plan for a manufacturing line producing 5-ohm resistors. We take a random sample of 10 resistors and calculate the sample average resistance $\bar{x}$. If $|\bar{x} - 5| \leq .5$, we accept the null hypothesis that $\mu = 5$. If $|\bar{x} - 5| > .5$, we accept the alternative hypothesis that $\mu \neq 5$.
   a. If $\sigma = 4$, calculate the level of significance; i.e., calculate $P(|\bar{x} - 5| > .5)$ given that $\mu = 5$ and $\sigma = 4$.
   b. With $\sigma = 4$, calculate the probability of accepting the null hypothesis when $\mu = 6$; i.e., calculate $P(|\bar{x} - 5| \leq .5)$ given that $\mu = 6$ and $\sigma = 4$.
5. Does $\alpha + \beta = 1$? Why or why not?
6. The keypunch operators in a computer center claim that at least 90% of cards punched are free of errors. A sceptical user of the center selects 10 cards at random from a day's output and finds 3 cards with

errors. Calculate the observed significance level for the null hypothesis that $p \leq .10$ versus the alternative hypothesis that $p \geq .10$, where p is the fraction of cards with errors.

7. A larger study than in Exercise 6 is made of the computer center. One hundred cards are selected at random, and 15 cards are found with errors. Using the central limit theorem, find the probability that $Z > Z_{\text{observed}}$ where

$$Z_{\text{observed}} = \frac{15 - 100p}{\sqrt{100p(1 - p)}}$$

and where $p = .10$. Do the data support the computer center claim?

8. If male and female births were equally likely, the Arbuthnot data of Section 17.1 would be very surprising. Consider a less surprising set of data. Suppose that in 82 years male births exceeded female births in only 62 years. Calculate the observed significance level for the null hypothesis that $p = \frac{1}{2}$ versus the alternative hypothesis that $p \neq \frac{1}{2}$. Use the Z statistic of Exercise 7 with $p = \frac{1}{2}$ and find $P(|Z| > Z_{\text{observed}})$.

9. Muffins baked from flour containing 15% peanut flour are given to a taste panel of 16 people. Each muffin is given a score on a five-point rating scale on each of 10 characteristics. Then the average score over all 10 characteristics is recorded for each muffin. The 16 scores are 3.1, 3.2, 3.1, 2.8, 2.9, 2.6, 3.0, 3.2, 3.7, 3.9, 3.3, 2.7, 3.2, 2.4, 2.8, and 3.3. Using $\alpha = .05$, test the null hypothesis that $\mu = 3.0$ (wheat flour) versus the alternative hypothesis that $\mu \neq 3.0$. Accept or reject the null hypothesis.

10. A particular bank offers several different accounts and assesses no charge on a checking account with a \$500 average daily balance. Since no interest is accrued on checking account balances, the supposition is that customers try to maintain small balances as close to the required balance as possible. A quick check of checking account balances gives the following 10 average daily balances: 491, 523, 486, 494, 532, 564, 488, 422, 562, and 541. Using $\alpha = .05$, test the null hypothesis that $\mu = 500$ against the alternative hypothesis that $\mu > 500$. Accept or reject the null hypothesis.

11. Student evaluations are made on every course at a certain university. In a large department the average instructor rating is about 3.25 year after year. In 20 upper-division sections the ratings are 3.05, 3.31, 3.34, 3.82, 3.30, 3.16, 2.84, 3.10, 2.90, 3.18, 2.88, 3.22, 3.28, 3.34, 3.62, 3.28, 3.30, 3.22, 3.54, and 3.30. Calculate the observed

significance level for the hypothesis that $\mu = 3.25$ against the alternative that $\mu > 3.25$. Do the data give strong support to the idea that upper-division ratings are kinder to the instructors?

12. The people in a neighborhood made a count of the number of children coming to their houses for trick or treat on Halloween night. They believed that some children were coming from across town to their neighborhood. The numbers counted were 88, 112, 96, 92, 93, 98, 89, 101, 86, 94, 95, 68, 72, 102, 94, 81, 84, 85, 92, 87, 76, and 89. Calculate the observed significance level for a mean of 84 against a hypothesis of a greater mean.
13. The state coordinating board for higher education in a certain state tries to monitor enrollment trends. The changes in enrollment from the previous year are reported by 24 colleges. The changes are 224, 208, -111, 362, 15, -76, -282, -106, 104, 172, -108, -222, 500, 46, 215, 178, 337, 206, 79, 42, 39, 162, -317, and -226. Do the data indicate a decrease or an increase in the true enrollment? Use a t test to test the null hypothesis of a zero mean versus $\mu \neq 0$. Use $\alpha = 0.5$.
14. An entomology researcher is constructing a mathematical model to describe the impact of boll weevils on cotton plants. She has measured the number of blooms per cotton plant on each day of the growing season. On one particular day the numbers are 42, 26, 15, 19, 27, 32, 26, 24, 17, 19, 22, 18, 24, 16, 14, 30, 22, 25, 27, and 29. She wishes to use 20 as the mean number of blooms on this particular day. Is this a reasonable number? Base your answer on a t test.
15. A horticulturist has conducted a greenhouse experiment to study the effect of watering rates and feeding rates on chrysanthemum plants. At the end of the season, he measures the diameter of the largest bloom on each plant. The diameters, in inches, are $6\frac{1}{4}$, $6\frac{1}{8}$, $5\frac{7}{8}$, $8\frac{3}{4}$, $7\frac{3}{16}$, $8\frac{1}{8}$, $7\frac{7}{16}$, $5\frac{15}{16}$, $6\frac{3}{8}$, $6\frac{1}{4}$, $6\frac{1}{2}$, $7\frac{1}{4}$, $8\frac{1}{2}$, and $7\frac{1}{8}$. Calculate the observed significance level for the hypothesis that the mean diameter of the largest bloom is 6 inches against the alternative of a larger diameter.

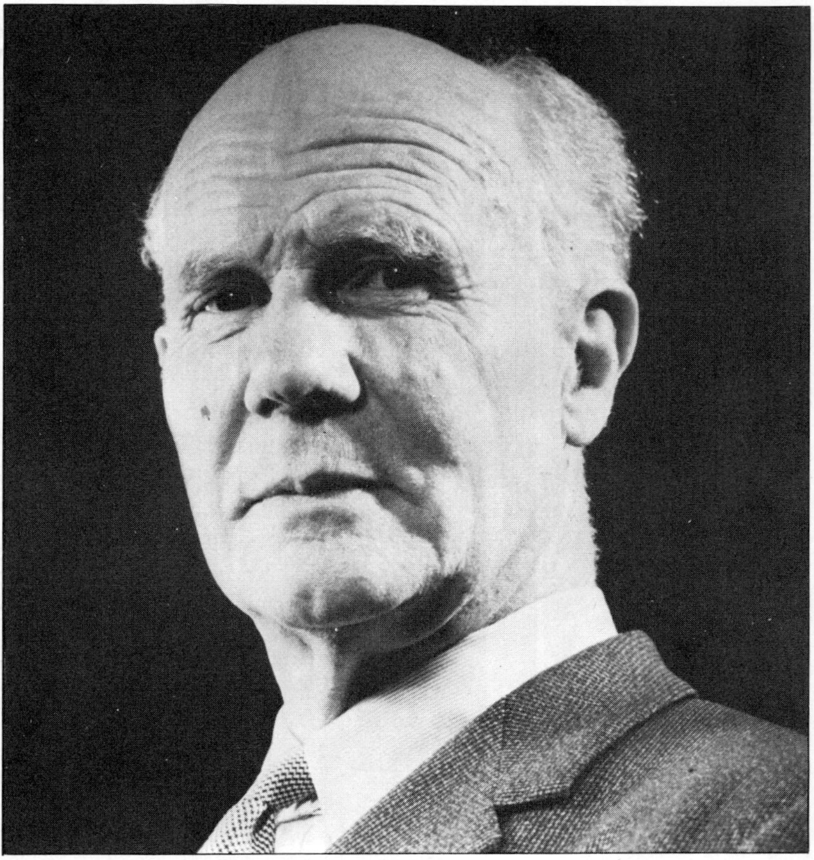

E. S. PEARSON. 1895–1980. Keystone Press Agency.

18

ESTIMATION

18.1 *Tuppence patiently, cautiously, trustingly invested in the, to be specific, in the Dawes-Toomes-Mousley-Grooms Fidelity Fiduciary Bank.**

Mary Poppins

Although Jane and Michael Banks were not convinced by Mr. Dawes of the desirability of investing their tuppence in the bank, millions of people do place their **trust** and **confidence** in banks and other **fiduciary** establishments. The words **fiducial** and fiduciary derive from the Latin word **fiducia**, for trust. In this lecture we discuss the estimation of parameters by methods that we consider to be trustworthy.

18.2 In an earlier section we referred to **point estimation** as the process of calculating specific values from sample data to be used for values of parameters occurring in the population probability distribution. The development of a theory of point estimation began essentially with the work of Fisher in the early 1920s. In two major papers Fisher [1, 2] clearly stated the problem of estimation and developed several methods of measuring the **goodness** of an estimate. Phrased another way, he developed several methods of measuring our faith, or confidence, in an estimate. We will discuss only two properties of estimates: **unbiasedness** and **minimum variance**.

18.3 In an earlier section we demonstrated that the mean of the sampling distribution of $\bar{x}$ is the mean of the parent population, μ. This property of $\bar{x}$ is underscored by saying that $\bar{x}$ is an **unbiased estimate** of μ. As used in

* "Fidelity Fiduciary Bank," copyright © 1963 Wonderland Music Company, Inc. Words and Music by Richard M. and Robert B. Sherman.

this technical sense, unbiasedness has nothing to do with fairness or lack of prejudice. It simply means, as already stated, that the average value of all possible $\bar{x}$s is the value of μ, which we are estimating, or, in symbols,

$$\text{Ave}(\bar{x}) = \mu$$

Another way of verbalizing this property is to say that $\bar{x}$ will give us the correct value for μ on the average.

In more general notation, we will say that $\hat{\theta}$ (theta hat) is an unbiased estimate of θ if the average value of all possible values of $\hat{\theta}$ is equal to θ or if, in symbols,

$$\text{Ave}(\hat{\theta}) = \theta$$

As another example of an unbiased estimate, consider the sample variance s^2. We demonstrated earlier, in sampling from an infinite population, that the average value of all possible values of s^2 is equal to σ^2, the variance of the parent population. So we say that s^2 is an **unbiased estimate** of σ^2. To challenge students who are beginning to feel that they understand the idea, we would like to point out that s is **not** an unbiased estimate of σ.

Suppose we wish to estimate the average length of a queue at a bus stop. We might select times at random during a week and count the number of people standing in line for the bus at these times; we would obtain the following data: 10, 2, 15, 23, 7, 0, 5, 3, 8, 13. From these sample data, we wish to estimate the parameter θ, the average queue length during the week. How should we do this? I have asked this question of several classes of beginning students. Invariably, most popular answers are to estimate θ either by $\hat{\theta}_1$ = sample mean = 8.6 or by $\hat{\theta}_2 = 6.3$ (sample mean after deleting the suspiciously large number 23). The sample **median**, $\hat{\theta}_3 = 7.5$, is also usually suggested.

Without further assumptions about the frequency distribution of the population, we can claim unbiasedness for $\hat{\theta}_1$, the sample mean, only. If we assume that the distribution is symmetric about θ, both $\hat{\theta}_1$ and $\hat{\theta}_3$ are unbiased. Although $\hat{\theta}_2$ seems to be a reasonable ad hoc way to estimate θ, it will generally be biased downward.

18.4 Given several unbiased estimates of a parameter θ, we need some other criterion for comparing them and choosing between them. Consider the sampling distributions of $\hat{\theta}_1$ and $\hat{\theta}_3$ given that the population distribution is symmetric about θ (e.g., normal). We could depict these sampling distributions as in Figure 18.1. Although both estimates are centered on the true value, the sampling distribution of $\hat{\theta}_1$ is more concentrated and would be preferred.

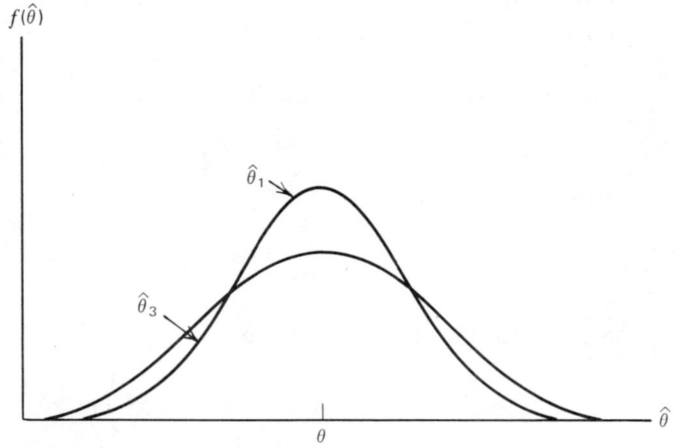

Figure 18.1 Comparison of Estimates.

Given two or more unbiased estimates, an obvious way to compare them is by comparing their variances. The estimate with the smallest variance will generally have a sampling distribution that is most concentrated about θ and is, therefore, preferred. An estimate with the smallest variance of all possible unbiased estimates is called a **minimum variance unbiased estimate**, or sometimes a **best** estimate. The word best should be used advisedly. Sometimes a novice may misinterpret it to mean that the estimate is best in all sorts of ways, other than having minimum variance among the unbiased estimates.

The sample mean is undoubtedly the most common way of estimating the population mean. Regardless of the form of the population distribution, the sample mean is unbiased. Furthermore, if the population is normal, the sample mean is a minimum variance unbiased estimate.

For the bus queue example, it is probably unlikely that the population is normal or even symmetric. Instead, the distribution is likely to be **skewed** to the right, with many small numbers and a few very large numbers. In such a case, the sample mean is unbiased but not minimum variance unbiased. Without further assumptions, we cannot say what the minimum variance unbiased estimate is.

18.5 A very natural way of expressing belief in an estimate is to give an **interval** within which we are reasonably certain that a population parameter lies. For example, engineers commonly add and subtract an error of 10% as an arbitrary safety factor. We now want to go through a simple probability argument and give several different interpretations that have flowed from it. Suppose that we have a random sample of size n from a normal population with mean μ and variance σ^2, with σ known but μ unknown. Now we can obtain from the standard normal table that

$$P(-1.96 \leq Z \leq 1.96) = .95$$

Also, we know that

$$\frac{\bar{x} - \mu}{\sigma/\sqrt{n}}$$

is a standard normal variable, so

$$P\left(1.96 \leq \frac{\bar{x} - \mu}{\sigma/\sqrt{n}} \leq 1.96\right) = .95$$

Multiplying all terms inside the parentheses by $\sigma/\sqrt{n}$, we obtain

$$P(-1.96\sigma/\sqrt{n} \leq \bar{x} - \mu \leq 1.96\sigma/\sqrt{n}) = .95$$

Next, we subtract $\bar{x}$ from each term to obtain

$$P(-\bar{x} - 1.96\sigma/\sqrt{n} \leq -\mu \leq -\bar{x} + 1.96\sigma/\sqrt{n}) = .95$$

Finally, we multiply all terms by -1, reversing inequalities, to obtain

$$P(\bar{x} - 1.96\sigma/\sqrt{n} \leq \mu \leq \bar{x} + 1.96\sigma/\sqrt{n}) = .95$$

Thus we obtain the interval from $\bar{x} - 1.96\sigma/\sqrt{n}$ to $\bar{x} + 1.96\sigma/\sqrt{n}$ as an estimate of μ.

18.6 Fisher published a paper called "Inverse Probability" in 1930 [3] in which he introduced the idea of **fiducial probability**, which resulted in **fiducial intervals**. At about the same time, Neyman developed the idea of **confidence intervals**. The idea of confidence intervals first appeared in print in 1932 in a booklet by Pytkowski [5]. For the next 10 years or so, numerous books and articles described confidence intervals and fiducial intervals as being the same in concept but differently named. Gradually, as Fisher and Neyman developed their ideas, it became clear that fiducial intervals and confidence intervals were quite different.

The fiducial interval interpretation for the interval $\bar{x} \pm 1.96\sigma/\sqrt{n}$ obtained in Section 18.5 can be summarized as follows.

1. The probability is for the **specific** value of $\bar{x}$ obtained in the sample.
2. The probability statement is made **after** the sample is taken and $\bar{x}$ is calculated.
3. The probability is an expression of **belief** about the value of μ, a fixed parameter.

By way of contrast, the confidence interval interpretation for the interval is as follows.

1. The probability is **not** for the specific value of $\bar{x}$ obtained in the sample.
2. The probability statement is made **before** the sample is taken and $\bar{x}$ is calculated.
3. The probability is the **relative frequency** with which intervals calculated using the formula will contain the parameter μ.

It is clear, as far as this example is concerned, that the difference is a difference of interpretation. It results from the two different ways of thinking about probability that we have mentioned previously: probability as a measure of belief and probability as a property of a physical system.

Fisher's concept of fiducial probability was not generally accepted. Instead, Neyman's confidence interval theory became extremely well known and is very widely taught.

18.7 In recent years **Bayesian confidence intervals** have been increasing in favor. As far as the example of Section 18.5 is concerned, the Bayesian interpretation is similar to the fiducial interpretation. The probability interpretation is that of degree of belief and 95% confidence for the interval on μ from $\bar{x} - 1.96\sigma/\sqrt{n}$ to $\bar{x} + 1.96\sigma/\sqrt{n}$ is interpreted as a measure of the belief attached to the statement that μ lies in the interval. Although the interpretation of probability is similar for Bayesian intervals and fiducial intervals, the method of obtaining the intervals is quite different. Bayesian intervals are obtained by assigning probabilities to μ before the sample is drawn and modifying these probabilities using Bayes' theorem after the sample is drawn.

18.8 A slightly different way of obtaining intervals of values supported by the data is given by Kempthorne and Folks [4]. Proceeding from a test of a null hypothesis, the significance level, SL, is graphed versus possible hypothesized values for μ. For example, suppose we draw a random sample of size $n = 25$ from a normal distribution with unknown mean μ and variance $\sigma^2 = 16$. With $H_0: \mu = \mu_0$ versus $H_A: \mu \neq \mu_0$, we would evaluate the significance level by finding the probability of a standard normal variable exceeding $\sqrt{n}(\bar{x} - \mu_0)/\sigma$, or, by

$$SL = P[|Z| > |\sqrt{n}(\bar{x} - \mu_0)/\sigma|]$$

Given results of the sample, we can evaluate the significance level for any specified sample. Suppose we obtained $\bar{x} = 10$. Then for

$$\mu_0 = 10, \qquad \text{SL} = P(|Z| > 0) = 1$$

for

$$\begin{aligned} \mu_0 = 12, \qquad \text{SL} &= P[|Z| > |\sqrt{25}(10 - 12)/4|] \\ &= P(|Z| > 2.5) \\ &= .0142 \end{aligned}$$

and for

$$\begin{aligned} \mu_0 = 9, \qquad \text{SL} &= P[|Z| > |\sqrt{25}(10 - 9)/4|] \\ &= P(|Z| > 1.25) \\ &= .2112 \end{aligned}$$

Proceeding in this way, we can plot the significance level versus μ_0 in Figure 8.2.

Given any significance level, we can obtain the set of μ values in agreement with (in consonance with) the data. These intervals were called **consonance intervals** by Kempthorne and Folks.

The intervals obtained are exactly the same as confidence intervals, but the interpretation is different. In any case, the graph in Figure 18.2 is worth constructing.

18.9 In this lecture we have presented some of the underlying concepts of **point** and **interval estimation**. The ideas may seem somewhat abstract to beginning statistics students who are not accustomed to considering alternative estimates. Yet it is evident that in matters of national concern

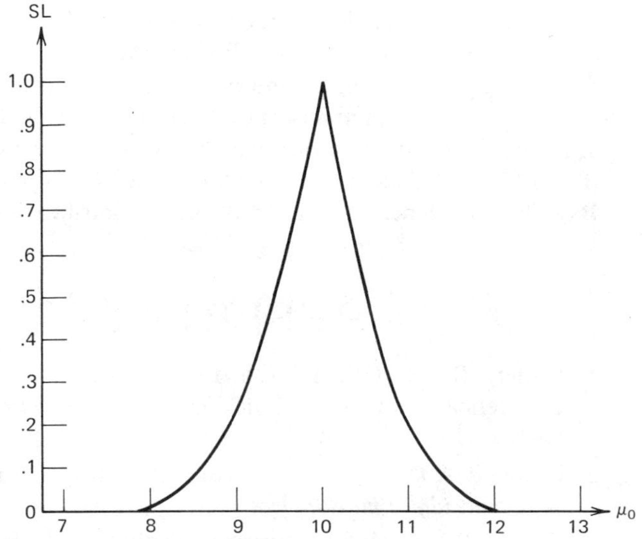

Figure 18.2 Consonance Intervals.

(air pollution, nuclear radiation, energy supply, etc.), many alternative estimates are feasible from the same set of data. We have tried to emphasize in this lecture that the goodness of an estimation procedure cannot be judged from its performance on a single set of data. Instead, the goodness of the procedure must be based on its properties within a broader context. With the relative frequency interpretation of probability, the performance for many possible samples is studied. With a measure of belief interpretation, the performance is judged on the basis of a broader probability model. In both cases we must consider more than the single set of data at hand.

SUMMARY. Point estimation is the process of calculating a specific value, called an **estimate**, from sample data to be used for the parameter value in a model. The goodness of an estimate cannot be judged from a single sample; it is judged from the behavior of the method of estimation in the totality of all possible samples.

An **unbiased** estimate is obtained by a procedure such that the average of all possible values equals the parameter being estimated. In the class of all unbiased estimates, the one with the smallest variance is called the **minimum variance unbiased estimate**.

Intervals of plausible parameter values are sometimes called **interval estimates**. In the early 1930s Fisher and Neyman put forward independent formulations of interval estimates. Fisher's **fiducial interval** carried with it a probability statement about the parameter value for a specified sample. By contrast, Neyman's **confidence** interval carried with it a probability statement about intervals for a specified parameter. **Bayesian confidence intervals** are similar in spirit to fiducial intervals but are obtained by the use of Bayes' theorem.

REFERENCES

1. Fisher, R. A. (1922). "On the Mathematical Foundations of Theoretical Statistics," *Philosophical Transactions of the Royal Society, A-222*, 309–368.
2. Fisher, R. A. (1925). "Theory of Statistical Estimation," *Proceedings of the Cambridge Philosophical Society, 22*, 700–725.
3. Fisher, R. A. (1930). "Inverse Probability," *Proceedings of the Cambridge Philosophical Society, 26*, 528–535.
4. Kempthorne, O., and Leroy Folks. (1971). *Probability, Statistics, and Data Analysis*, Ames: Iowa State Press.
5. Pytkowski, W. (1932). The dependence of income of small farms upon their area, the outlay, and the capital investment in cows, Warsaw: *Series Bibljoteka Pulawska, 34*.

EXERCISES

1. Choose a random sample of the students in your class and obtain an unbiased estimate of the average height of all the students in the class.
2. Let t_0 be a tabulated "Student's" t value such that

$$P(-t_0 < t < t_0) = \text{desired confidence}$$

For example, with 15 degrees of freedom and a desired confidence of 0.95, $t_0 - 2.131$. Starting with

$$P\left(-t_0 < \frac{\bar{x} - \mu}{s/\sqrt{n}} < t_0\right) = \text{desired confidence}$$

go through steps analogous to those in Section 18.5 to obtain the confidence interval formula

$$\bar{x} \pm t_0 s/\sqrt{n}$$

3. Obtain a 95% confidence interval for the mean class height using the data collected in Exercise 1.
4. Gardeners are always interested in the average frost-free date for their locality. In a certain city, inspection of the weather records gave the last killing frost of the year for 20 years.

March 10	April 15	April 13	April 22	March 28
April 24	May 6	April 5	April 18	May 2
April 5	April 9	March 26	April 14	April 17

   Assuming that the time of frost as measured from January 1 is normally distributed, obtain a 95% confidence interval for the average frost-free date.
5. The Environmental Protection Agency has established maximum allowable increases in levels for air pollutants. In a certain city the maximum allowable 24-hour increase was 91 micrograms per cubic meter. Measurements at ten randomly selected sites in the city showed the following 24-hour increases: 89, 92, 84, 60, 76, 94, 78, 79, 74, 81. Obtain a 95% confidence interval for the 24-hour increase. Do you think the city was in violation of the standard?
6. This exercise illustrates the concept of unbiasedness. Consider the finite population 4, 2, 15, 1, 8. List all 10 of the possible samples of size three that can be obtained in sampling without replacement. For each possible sample, calculate the sample mean, the sample median and the sample variance s^2.
   a. Show that the sample mean is an unbiased estimate of the population mean by showing that the average of the 10 $\bar{x}$s equals the mean of the population.
   b. Similarly, show that the sample s^2 is an unbiased estimate of the population S^2.
   c. Similarly, show that the sample median is not an unbiased estimate of the population mean.
   d. Show that the sample s is not an unbiased estimate of the population S.
7. The utility company in a small town launched a campaign to help people conserve energy. Answering calls to conduct energy surveys of homes, a team from the utility company noted the thermostat settings. They were (°F) 78, 69, 72, 70, 67, 68, 69, 71, 66, 68, 70, 65, 71, 67, 68, 72, 69, 73, 64, and 70. Calculate a 90% confidence interval for the mean thermostat setting.
8. A statewide tax reform question was being submitted to the voters and a local survey was conducted to get an indication of voter

preferences. Out of a random sample of 50 people, 29 favored the proposal and 21 opposed it. Estimate the fraction favoring the question and calculate an approximate 95% confidence interval for the fraction favoring the question.

9. A tire manufacturer is interested in the slump in sales in a particular tire model and surveys people who have purchased tires of that model in the last 4 years. Those who used the tires until replacing them by another model were asked about the miles of service. The miles of service reported were:

22,325	24,320	18,260	20,200
24,350	25,000	19,700	22,500
23,250	24,000	22,000	21,900

Calculate a 95% confidence interval for the mean number of miles of service.

10. A cabinetmaker is using a new glue in a critical support joint in a kitchen cabinet and conducts some tests to assure himself of the strength of the glue joint. He glues two pieces of wood together and then uses weights to break the glue joint. The weight (pounds) required to break 10 glue joints is 545, 495, 625, 530, 585, 615, 580, 615, 590, and 540. Calculate a 95% confidence interval on the mean weight required to break the glue joints.

11. Ten pages were selected at random from a city telephone directory. The number of names on these 10 pages were 399, 412, 388, 422, 396, 420, 410, 406, 392, and 395. Calculate a 95% confidence interval for the number of names listed in the 900-page directory.

12. Estimate the fraction of printed material in this book devoted to exercises, calculate a confidence interval, and explain your methods.

TWO DIMENSIONS

ARTILLERY. Firing errors—one of the first examples of the bivariate normal. The New York Public Library Picture Collection.

19

THE BIVARIATE NORMAL DISTRIBUTION

19.1 *Great things are done when men and mountains meet;
this is not done by jostling in the street.*

William Blake

Look at the **contour** map of Stone Mountain, Georgia (Figure 19.1). It is not difficult to envision a smoothly rounded surface and a steep slope on the north side of the mountain. Although not as informative as a three-dimensional model, a contour map helps us to visualize the surface of the earth. Thinking of the earth's surface with maps thereof provides a useful way of describing **bivariate** data, data that arise from populations where each member exhibits two variables.

19.2 **Bivariate frequency distributions** are useful in describing bivariate populations. Table 19.1 is one of many bivariate frequency distributions given by Pearson and Lee [2]. Although the entries in the table are basically frequencies, they require some explanation because of fractional values. The data were originally recorded to the nearest inch intervals. When an individual fell on the boundary of a cell, its frequency was split between the cells. To illustrate the idea, consider the three father-son pairs in Table 19.2. The frequency of one for the father-son pair (68.5, 76.5) is split by giving 0.25 frequency to the four corner cells, producing Table 19.3. These fractional frequencies are, in fact, the values recorded in Table 19.1.

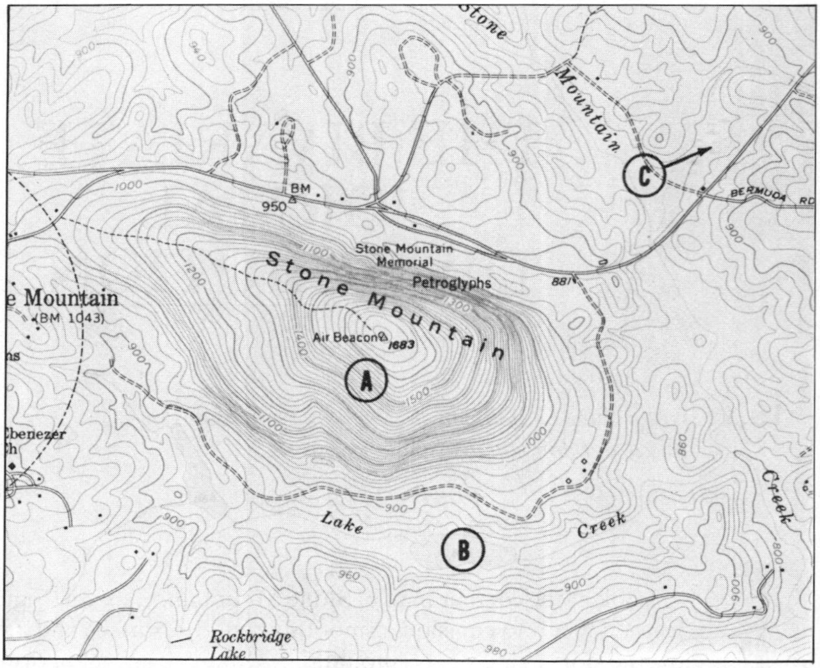

Figure 19.1 Stone Mountain. (Reproduced from Figure V-8, *Atlas of Landforms* by J. L. Scovel et al. Copyright © 1966 by John Wiley & Sons, Inc. Also reprinted from the U.S. Geological Survey. Reprinted by permission from John Wiley & Sons, Inc.)

19.3 Imagine erecting at the center of each cell in Table 19.1 a vertical line proportional to the frequency of that cell. If the tops of the vertical lines are connected, we have a surface like the surface of a mountain. The beautiful drawing in Figure 19.2 is given by Yule and Kendall [5].

19.4 Consider describing a population as we have just done, not for 1078 father-son pairs but for all of the millions of father-son pairs in the United States. Furthermore, imagine that instead of tabulating heights to the nearest inch, we tabulate to the nearest one-tenth of an inch, etc. For such bivariate frequency distributions, we construct three-dimensional pictures like that in Figure 19.2. As the measurements become more precise, the surface depicted will become smoother, and we can visualize the limiting surface to be like that in Figure 19.3.

Table 19.1 Correlation between (1) Stature of Father and (2) Stature of Son: 1 or 2 Sons only of each Father. Measurements in inches.

(2) Stature of Son.	(1) Stature of Father. 58·5–59·5.	59·5–60·5.	60·5–61·5.	61·5–62·5.	62·5–63·5.	63·5–64·5.	64·5–65·5.	65·5–66·5.	66·5–67·5.	67·5–68·5.	68·5–69·5.	69·5–70·5.	70·5–71·5.	71·5–72·5.	72·5–73·5.	73·5–74·5.	74·5–75·5.	Total.
59·5–60·5	—	—	—	—	·5	·5	1	—	—	—	—	—	—	—	—	—	—	2
60·5–61·5	—	—	—	—	·5	—	—	—	1	—	—	—	—	—	—	—	—	1·5
61·5–62·5	—	·25	·25	—	·5	1	·25	·25	·5	·5	—	—	—	—	—	—	—	3·5
62·5–63·5	—	·25	·25	2·25	2·25	2	4	5	2·75	1·25	—	·25	·25	—	—	—	—	20·5
63·5–64·5	1	—	1·5	3·75	3	4·25	8	9·25	3	1·25	1·5	·75	1·25	—	—	—	—	38·5
64·5–65·5	2	1	·5	2	3·25	9·5	13·5	10·75	7·5	5·5	3·5	2·5	—	—	—	—	—	61·5
65·5–66·5	—	·5	1	2·25	5·25	9·5	10	16·75	17·5	16	5·25	2	2·5	1	—	—	—	89·5
66·5–67·5	—	1·5	2	4·75	3·5	13·75	19·75	26·5	25·75	19·5	12·5	13·75	3·25	·5	1	—	—	148
67·5–68·5	—	—	1·5	2	7·5	10	10·25	24·25	31·5	23·5	29·5	13·25	8·5	9·5	2·25	—	—	173·5
68·5–69·5	—	—	1	—	5·25	5	12·75	18·25	16	24	29	21·5	10	3·5	2·25	—	1	149·5
69·5–70·5	—	—	—	—	1	2·5	5·75	18·75	11·75	19·5	22·5	19·5	14·5	6·25	3·5	1·5	1	128
70·5–71·5	—	—	—	—	—	3·25	5	8·75	10·75	19	14·75	20·75	10·75	8	5	1	1	108
71·5–72·5	—	—	—	—	—	·25	3	1·25	7	7·75	10·75	11·25	10	8·5	2·75	·5	—	63
72·5–73·5	—	—	—	—	—	—	·75	·75	2·5	7·5	6·5	6	7·5	6·25	3·25	·5	·5	42
73·5–74·5	—	—	—	—	1	—	1·5	1·5	—	5·25	2·25	2·5	6·5	3·25	3·25	—	2	29
74·5–75·5	—	—	—	—	—	—	—	—	—	1	2	—	2·5	·75	1·75	·5	—	8·5
75·5–76·5	—	—	—	—	—	—	—	—	—	1·25	·25	—	·5	1	1	—	—	4
76·5–77·5	—	—	—	—	—	—	—	—	—	1·25	·25	1	—	—	1·5	—	—	4
77·5–78·5	—	—	—	—	—	—	—	—	—	—	1	1	—	·25	·75	—	—	3
78·5–79·5	—	—	—	—	—	—	—	—	—	—	—	—	—	·25	·25	—	—	·5
Total	3	3·5	8	17	33·5	61·5	95·5	142	137·5	154	141·5	116	78	49	28·5	4	5·5	1078

Source. Reproduced from *Biometrika, 2,* Karl Pearson and Alice Lee, On the laws of natural inheritance in man. Also reproduced from Table 11.3, *An Introduction to the Theory of Statistics,* 11th edition, by G. U. Yule and M. G. Kendall.

Table 19.2 Frequencies

Son	Father 67.5–68.5	68.5	68.5–69.5
75.5–76.5	1		
76.5		1	
76.5–77.5	1		

Table 19.3 Splitting Frequencies

Son	Father 67.5–68.5	68.5–69.5
75.5–76.5	1.25	0.25
76.5–77.5	1.25	0.25

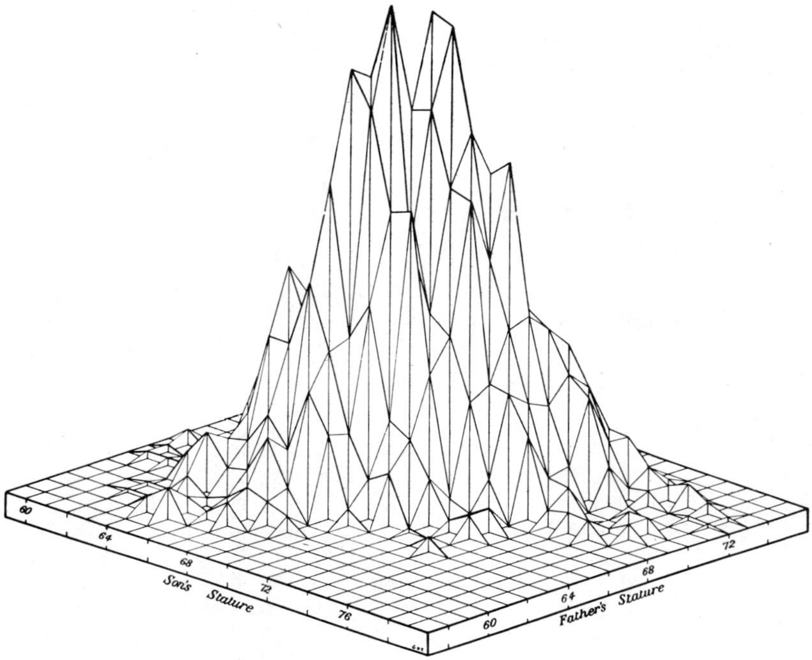

Figure 19.2 Frequency-Surface for Father-Son Data. (Reproduced from Figure 11.3, *Introduction to the Theory of Statistics,* 11th edition, by G. U. Yule and M. G. Kendall. Copyright © 1937 by Charles Griffin & Company Ltd. Reprinted by permission from Charles Griffin & Company Ltd.)

Many bivariate data sets are well described by surfaces like that in Figure 19.3 and are said to be bivariate normal. Of course, we may prefer to describe such surfaces by contour maps. These maps usually will look like Figure 19.4, which consists of **concentric ellipses**.

We will not present the mathematical equation for the bivariate normal. For the present we want to point out the following properties.

1. The bivariate normal distribution surface is a bell-shaped **surface** that is symmetric about the means of both variables.
2. If all of the frequencies are shoved to one axis, a **univariate** normal curve is obtained. For example, in the father-son data the stature of fathers only is obtained by ignoring the height of the sons and shoving all frequencies to the margin. A normal distribution for height of fathers is obtained.
3. The cross section along any plane perpendicular to the axis for either variable is also a normal bell-shaped curve.

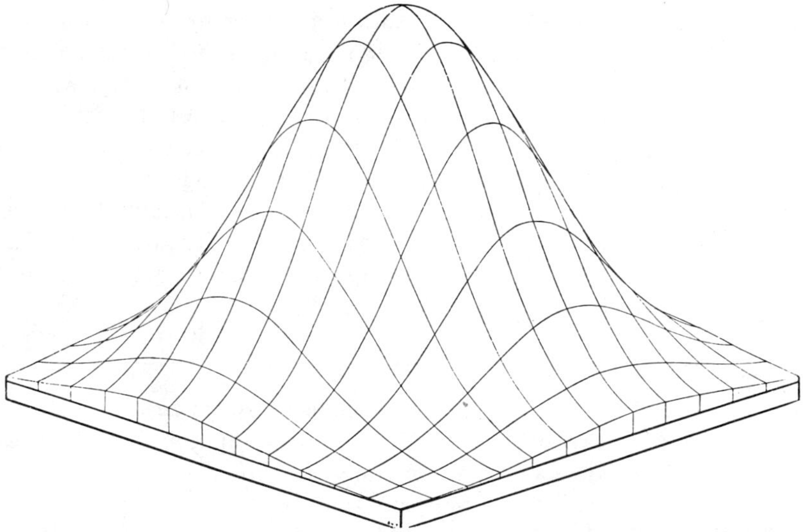

Figure 19.3 Limiting Form of the Frequency Surface. (Reproduced from Figure 11.1, *Introduction to the Theory of Statistics,* 11th edition, by G. U. Yule and M. G. Kendall. Copyright © 1937 by Charles Griffin & Company, Ltd. Reprinted by permission from Charles Griffin & Company Ltd.)

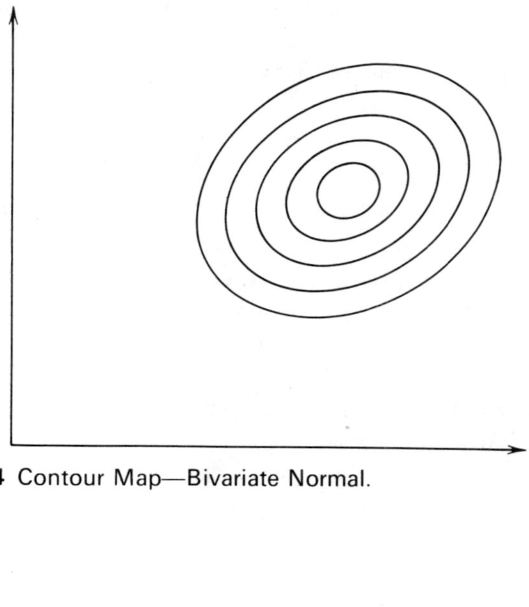

Figure 19.4 Contour Map—Bivariate Normal.

Walker [4] credits Adrain as being the first writer to mention a bivariate normal distribution as early as 1808. Walker also cites Laplace (1810), Plana (1812), Gauss (1823), and Bravais (1846) with arriving at an equation for what we now call the bivariate normal distribution. Pearson [1] credits Galton (1885) as being the first to think of the bivariate normal surface as describing observational variables. However, Schols (1875) discussed the bivariate normal as describing horizontal and vertical errors in artillery fire. Galton's work is of enormous importance because it represents a major beginning in the use of the bivariate normal in statistics.

SUMMARY. The use of **contour maps** to describe mountainous land surfaces provides a useful analogy for thinking about the distribution of **bivariate data**, data obtained by observing two variables for each member of the population. Classes, class intervals, and class marks are decided on for each variable. **Bivariate frequency distributions** are formed by recording the frequency for each pair of classes.

A three-dimensional graph of the bivariate frequency distribution can be formed by plotting the frequency for each two-way class as the ordinate value. Then the population **surface** is easily visualized.

The **bivariate normal distribution** has a bell-shaped surface. Furthermore, if the frequencies are considered on either margin, the **marginal distribution** is a univariate normal. The conditional distribution along my cross-sectional parallel to either axis is also normal. The **contours** of the bivariate normal distribution are **concentric ellipses**.

The origin of the bivariate normal is somewhat obscure, but it was known at least as early as 1808.

REFERENCES

1. Pearson, Karl. (1920). "Notes on the History of Correlation," *Biometrika, 13*, 25–45.
2. Pearson, Karl, and Alice Lee. (1903). "On the Law of Natural Inheritance in Man," *Biometrika, 2*, 357–462.
3. Schols, C. M. (1875). Over de theorie der fouten in de ruimte em in het platte vak," *Verhandlingen van de Koninklijke Akademie ven Wetenschappen, 15* (English translation, 1958; White Sands Proving Ground, New Mexico).
4. Walker, Helen M. (1920). *Studies in the History of Statistical Methods*, Baltimore: Williams and Wilkins.

5. Yule, G. U., and M. G. Kendall. (1937). *Introduction to the Theory of Statistics*, 11th Edition, London and High Wycombe: Charles Griffin.

EXERCISES

1. A much better appreciation for figures such as Figure 19.2 usually results when one tries to construct a similar figure. Construct a frequency surface for the following small set of bivariate data on automobile engines.

Net Torque	Net Horsepower 105–108	108–111	111–114
195–200	5	2	0
200–205	8	7	2
205–210	4	10	6
210–215	0	5	9

2. Repeat the Pearson and Lee inquiry. Try to get the heights on 100 father-son pairs or 100 mother-daughter pairs. Construct a bivariate frequency distribution such as that in Table 19.1 and a frequency surface such as that in Figure 19.2.
3. Using freehand drawing methods, draw contour lines on Table 19.1 for frequencies 30, 25, 20, 15, and 10. Are the contour lines roughly elliptical in shape?
4. A university department is administering student evaluation of instructors and keeps a stock of No. 2 lead pencils. A statistics student notices that the shorter, more used pencils also have shorter erasers. Measurement of 500 pencils gives the following data.

Pencil Length (inch)	Eraser Length ($\frac{1}{16}$ inch) 0–1	1–2	2–3	3–4
2–3	16	9	5	5
3–4	12	42	13	8
4–5	36	88	37	10
5–6	7	30	92	16
6–7	6	8	40	20

Construct a graph of this frequency distribution. Draw contours on this frequency distribution.

5. A mail-order house analyzed the orders received from a large number of mail customers. For customers ordering both shirts and shoes the shirt size (sleeve length) and shoe size (length) were recorded.

	Shirt Size					
Shoe Size	32	33	34	35	36	38
$7\frac{1}{2}$–8	42	20	1	0	0	0
$8\frac{1}{2}$–9	40	86	42	17	0	0
$9\frac{1}{2}$–10	62	101	68	50	23	0
$10\frac{1}{2}$–11	15	76	125	53	43	1
$11\frac{1}{2}$–12	0	0	70	108	48	8
$12\frac{1}{2}$–13	0	0	5	69	86	30
$13\frac{1}{2}$–14	0	0	0	10	20	32

Graph the frequency distribution. Sketch contours for the frequency distribution.

6. The manager of a grocery supermarket is able to study all sorts of relationships with his new accounting system. A study of the total grocery bill versus the amount for meat gives the following distribution.

Meat Bill	Total Bill (dollars)				
(dollars)	0–20	20–40	40–60	60–80	80–100
0–10	50	42	37	12	6
10–20	25	35	52	50	60
20–30	0	11	40	56	53
30–40	0	1	22	32	39

Graph the frequency distribution.

FRANCIS GALTON. 1822–1911. Courtesy of the American Museum of Natural History.

20

CORRELATION

20.1 *At length, one morning, while waiting at a roadside station near Ramsgate for a train, and poring over the diagram in my notebook, it struck me that the lines of equal frequency ran in concentric ellipses. The cases were too few for certainty, but my eye, being accustomed to such things, satisfied me that I was approaching the solution.**

Galton, in his *Memories of My Life* [1, p. 302], gives this account of his being led to a family of **concentric ellipses** like those we encountered in the previous lecture. Galton goes on to tell about consulting J. D. Hamilton Dickson, tutor at St. Peter's College, Cambridge, to obtain the mathematical equation for the bivariate normal that would give rise to such concentric ellipses.

20.2 It is necessary, if we are to have anything of the understanding of Galton (nearly 100 years ago), to have a rudimentary knowledge of straight lines and ellipses.

The equation of a **straight line** (any straight line) has the form

$$Y = a + bX$$

where a and b are constants. The constants a and b have simple geometrical interpretations. The constant a is the ***Y*-intercept**, the

* Reproduced by permission of the publishers Methuen & Co. from *Memories of My Life* by Francis Galton.

distance from the origin at which the line intercepts the Y-axis. The constant b is the **slope** of the line where

$$\text{Slope} = \frac{\text{vertical difference between two points}}{\text{horizontal difference between two points}}$$

In Figure 20.1 four straight lines are shown. Students should verify that the slopes and intercepts are as given.

In Lecture 19 we noted that the contours of the father-son frequency data were concentric ellipses. It is extremely helpful in understanding the correlation coefficient to have a little knowledge of the equation of an ellipse. Most students realize that an ellipse is an oval, egg-shaped figure but that not just any oval will do. One interesting way to draw an ellipse is to attach two ends of a piece of string to a piece of paper (with some slack in the string), to take up the slack with a pencil, and to move the pencil while keeping the string tight. This is shown in Figure 20.2.

The two lines of symmetry (more commonly the parts of these lines within the curve) are called the **major** and **minor axes**. The major axis is the longer of the two segments; the minor axis is the shorter.

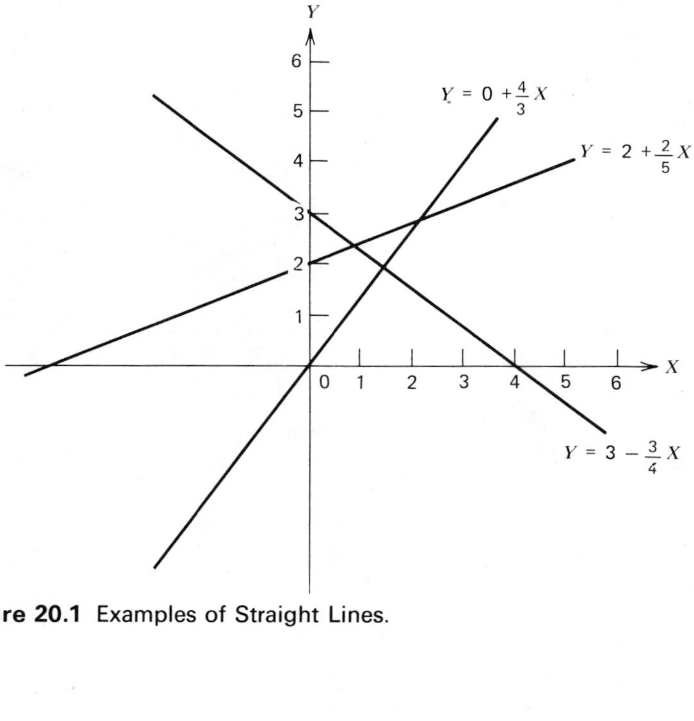

Figure 20.1 Examples of Straight Lines.

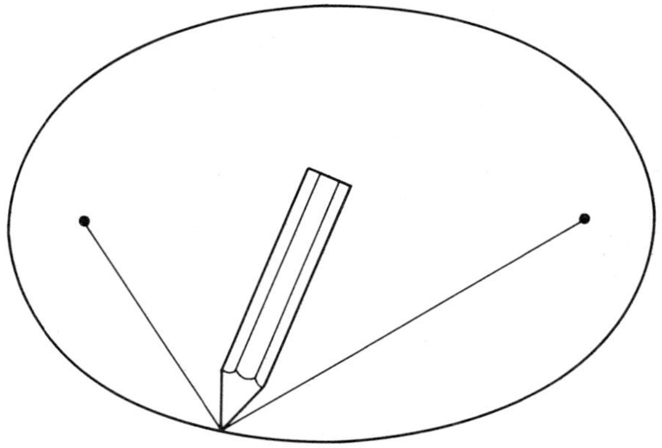

Figure 20.2 Construction of an Ellipse.

The general equation of any ellipse centered at the origin is given by

$$Ax^2 + 2Bxy + Cy^2 = D$$

with

$$AC - B^2 > 0$$

Specific ellipses are generated by assigning different values to A, B, C, and D. For example, if we let $A = 1$, $B = 0$, $C = 1$, and $D = 4$, we have a circle centered at the origin with radius 2.

20.3 If we now follow the step taken by Galton of **standardizing** our variables we can get a very simple measure of **association** or **correlation** between the standardized variables by examining the elliptical contours of the bivariate normal. Let the standardized variables be z_1 and z_2. Then these contours are described by the elliptical equation

$$z_1^2 - 2\rho z_1 z_2 + z_2^2 = D$$

where

$$1 - \rho^2 > 0$$

By varying the value for D, we get a family of ellipses concentric to the one in Figure 20.3.

If ρ is positive the major axis is the OB direction. If ρ is negative the major axis lies in the OA direction. If $\rho = 0$, the contours are circles. Thus ρ (rho), called the **correlation coefficient**, provides a measure of

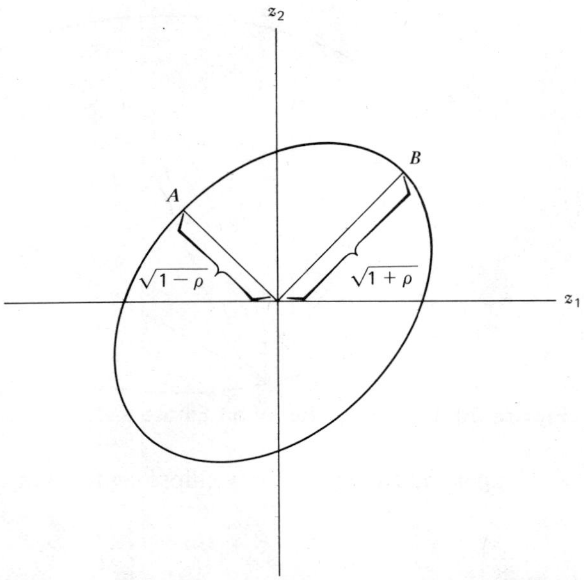

Figure 20.3 Typical Contour of Bivariate Normal: Standardized Variables.

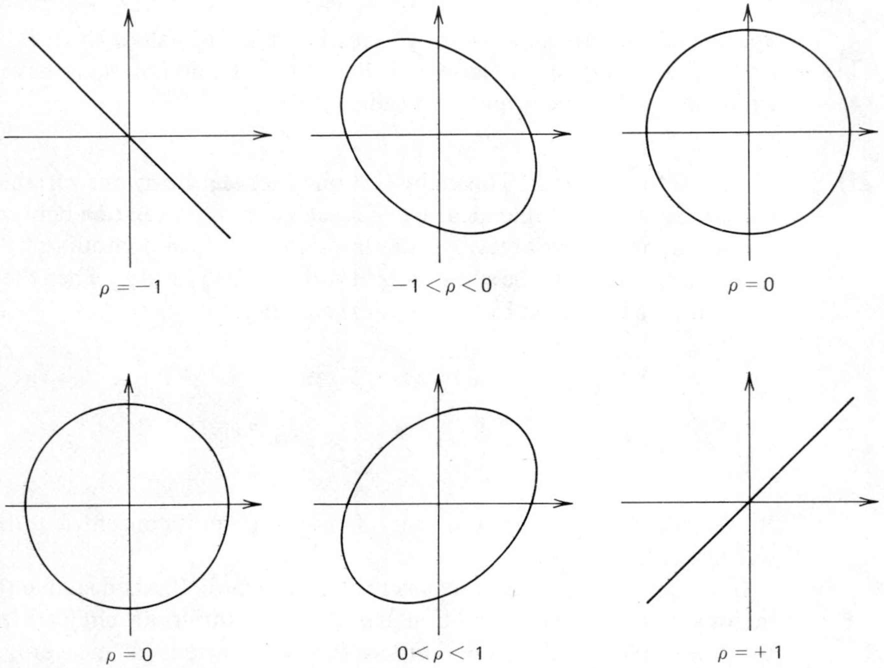

Figure 20.4 Interpretation of ρ.

association between the standardized variables z_1 and z_2 (and, therefore, between the unstandardized variables X and Y).

If $\rho > 0$, X will tend to be large when Y is large. If $\rho < 0$, X will tend to be large when Y is small. If ρ is close to $+1$ or -1, the relationship is very strong. The situation is depicted in Figure 20.4.

20.4 We have been discussing the correlation coefficient ρ as a parameter appearing in the equation for the bivariate distribution. The effect of ρ on the shape of the elliptical contours gives a way of thinking about the relationship between two normal variables X and Y. There is an alternative way of describing ρ that is equivalent to the one already given for bivariate normal populations and that is equally valid for nonnormal populations. Pearson [2] is generally credited with having extended the use of correlation to nonnormal populations, and this formulation is often called the **Pearson product-moment correlation.**

Consider a population in which a large X is usually associated with a large Y and a small X with a small Y. In terms of standardized variables z_1 and z_2, the scatter diagram would appear as in Figure 20.5.

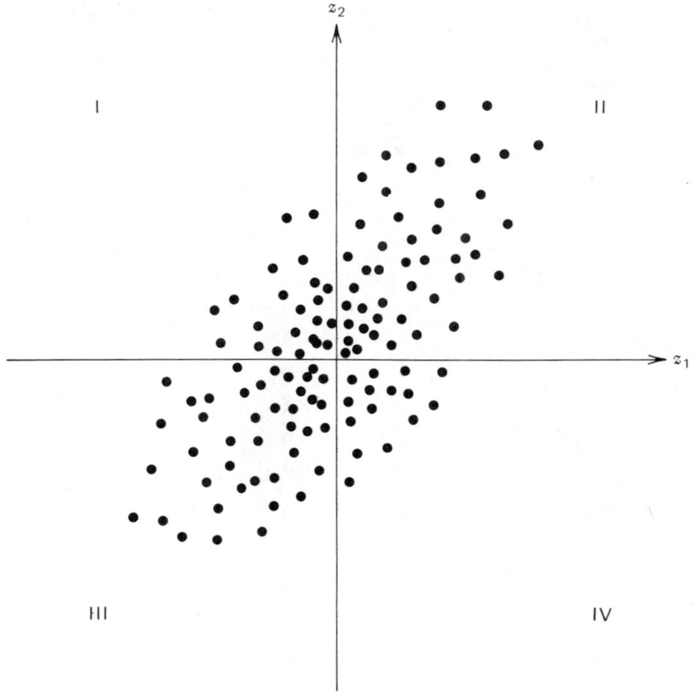

Figure 20.5 Product-Moment Correlation.

In quadrants I and III, the product $z_1 z_2$ would be positive; it would be negative in quadrants II and IV. Because there are more points in quadrants I and III than in quadrants II and IV, the average (or expected) product would be positive. If the scatter points tended to have a negative slope the average value of $z_1 z_2$ would be negative. It seems logical then to take this average product as a measure of association. It is shown in books on mathematical statistics that if we have a bivariate normal, this average product is identically equal to ρ as already defined. That is,

$$\begin{aligned}\rho &= \text{population average of } z_1 z_2 \\ &= \text{Ave}(z_1 z_2)\end{aligned}$$

In terms of the original variables

$$\rho = \frac{\text{Ave}(X - \mu_x)(Y - \mu_y)}{\sigma_x \sigma_y}$$

20.5 Given a sample, the most obvious way of estimating ρ is to calculate the sample equivalent. Denoting our estimate by r, let

$$\begin{aligned}r &= \frac{(1/n)\,\Sigma(X_i - \bar{X})(Y_i - \bar{Y})}{\sqrt{(1/n)\,\Sigma(X_i - \bar{X})^2\,(1/n)\,\Sigma(Y_i - \bar{Y})^2}} \\ &= \frac{\Sigma(X_i - \bar{X})(Y_i - \bar{Y})}{\sqrt{\Sigma(X_i - \bar{X})^2\,\Sigma(Y_i - \bar{Y})^2}}\end{aligned}$$

Example.

	X	Y	$X - \bar{X}$	$Y - \bar{Y}$	$(X - \bar{X})(Y - \bar{Y})$	$(X - \bar{X})^2$	$(Y - \bar{Y})^2$
	2	1	−0.4	−2.4	0.96	0.16	5.76
	3	6	0.6	2.6	1.56	0.36	6.76
	1	0	−1.4	−3.4	4.76	1.96	11.56
	4	8	1.6	4.6	7.36	2.56	21.16
	2	2	−0.4	−1.4	0.56	0.16	1.96
Total	12	17	0.0	0.0	15.20	5.20	47.20

$$\bar{X} = 2.4 \qquad \bar{Y} = 3.4$$

$$r = \frac{15.20}{\sqrt{(5.20)(47.20)}} = 0.9702$$

20.6 Assuming that X and Y are from a bivariate normal, it is very easy to test the null hypothesis

$$H_0: \rho = 0$$

versus the alternative hypothesis

$$H_1: \rho \neq 0$$

The sample correlation coefficient is simply calculated and compared with the value given in Table A.6.

Example. For the example in Section 20.5, $r = 0.9702$. From Table A.6, the tabled value for $n = 5$ is 0.878. Since the calculated value exceeds the tabled value, we conclude (at the 0.05 level) that $\rho \neq 0$.

SUMMARY. In searching for an explanation of association between two variables, Galton was led to the **correlation coefficient**. The correlation coefficient, ρ, has a very natural and simple explanation as a parameter of the equation generating the elliptical contours of the bivariate normal distribution. If X_1 and X_2 are closely related and tend to increase (or decrease) together, it seems reasonable that the contours are long, thin ellipses with a positive slope. The equation gives ρ a value close to $+1$. If X_1 and X_2 are closely related and X_1 tends to increase as X_2 decreases, the slope is negative and ρ is close to -1.

Pearson extended the use of correlation to nonnormal populations; for this reason the correlation coefficient is often called the **Pearson product-moment** correlation.

The sample correlation coefficient, r, is quite variable and inferences are suspect with small sample sizes. Testing of the hypothesis that $\rho = 0$ is very easy mechanically but requires a moderately large sample size to be of value.

REFERENCES

1. Galton, Francis. (1909). *Memories of My Life*, New York: E. P. Dutton.
2. Pearson, Karl. (1896). "Regression, Heredity, and Panmixia," *Philosophical Transactions, A-187*, 253–318.

EXERCISES

1. Graph and give equations for the following straight lines.
   a. Y-intercept $= 3$ slope $= 2$.
   b. Y-intercept $= -2$ slope $= 3$.
   c. Y-intercept $= -3$ slope $= -4$.
   d. Y-intercept $= -4$ slope $= -2$.

2. Sketch the ellipses for standardized variables z_1 and z_2 given by the equation

$$z_1^2 - 2\rho z_1 z_2 + z_2^2 = 1$$

for $\rho = -0.5, -0.2, 0.2$, and 0.5.

3. Television audiences in two neighboring cities noted that the weather forecasts for the two cities were often markedly different. However, their experience indicated that the weather in one city was remarkably similar to that of the other city. The following are midday temperature readings (°F) in the two cities over a 24-day period.

City X	City Y	City X	City Y
80.2	83.5	97.4	99.4
83.1	88.0	97.2	101.2
82.3	85.7	91.3	94.8
87.3	90.4	90.9	93.1
89.1	93.2	91.4	94.7
88.7	89.6	92.3	96.0
90.2	94.5	90.9	94.2
95.3	98.7	89.5	91.5
98.4	101.6	86.4	89.8
98.7	103.6	84.7	87.1
98.6	102.3	85.8	80.8
98.9	99.2	84.0	85.6

   a. Calculate the correlation coefficient r.
   b. Using Table A.6, test the hypothesis (at the 0.05 level) that $\rho = 0$.
   c. Plot the temperature readings on a scattergram.
   d. How would you describe the association between temperatures in the two cities?

4. Satellite data are used to estimate the acreage given to different kinds of land use. The percent of land planted to wheat estimated at 12 sites and estimates obtained from ground records were as follows.

Sites	Satellite	Ground
1	14	12
2	11	24
3	4	0
4	45	36
5	19	18
6	8	5
7	23	16
8	24	30
9	23	28
10	36	22
11	35	24
12	23	29

a. Calculate r.
b. Test $H_0: \rho = 0$ versus $H_1: \rho \neq 0$.

5. An interior decorator shops for brass candlesticks. It quickly becomes obvious that there is a relationship between the price and the reported age. We are given the following data for X = price in dollars and Y = age in years.

X	Y
85	5
125	150
165	170
140	150
120	125

Calculate the sample correlation coefficient, r, and test $H_0: \rho = 0$. Use $\alpha = .05$.

6. In a crowded classroom building the elevators receive heavy service. During the break between classes, there are people waiting to go up,

X, and people waiting to go down, Y. The building engineer collects some data on the numbers of people waiting. The data are as follows.

X	Y
32	28
16	18
48	36
52	35
36	28
38	39
43	40
52	34

Calculate the sample correlation coefficient, r, and test the hypothesis that $\rho = 0$. Use $\alpha = .05$. Are the results reasonable?

7. A salary survey in a particular profession asks for annual salary and age (among other variables). The data reported are as follows.

Age (X)	Salary (Y) (thousands of dollars)
30	22
41	28
52	38
38	40
48	34
56	35
44	32

Calculate the sample correlation coefficient, r, and test $H_0: \rho = 0$. Use $\alpha = .05$. Are salary and age related?

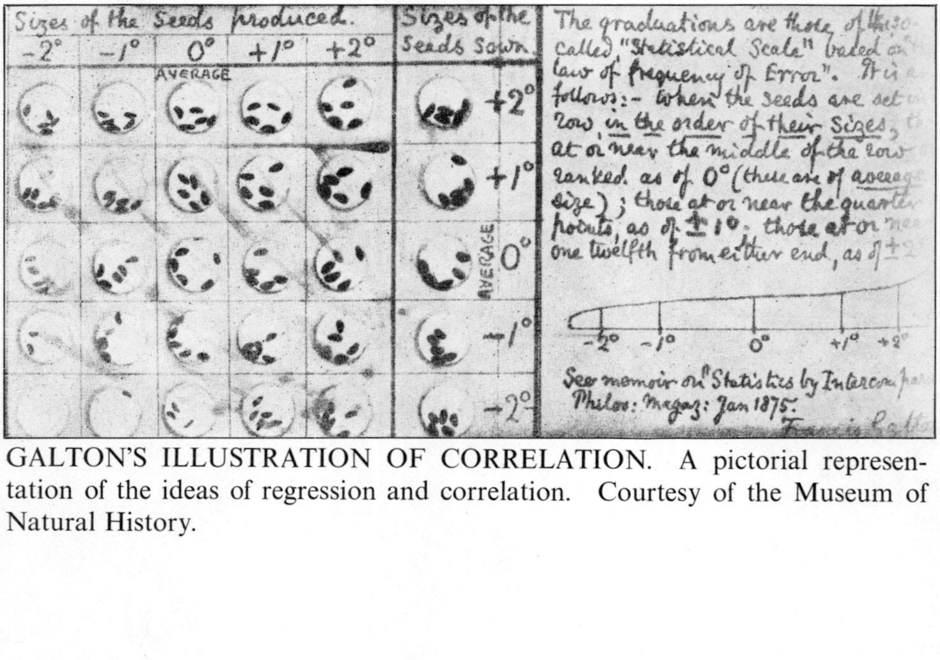

GALTON'S ILLUSTRATION OF CORRELATION. A pictorial representation of the ideas of regression and correlation. Courtesy of the Museum of Natural History.

21

POPULATION REGRESSION LINE

21.1 Forrest [1] gives a good account of the wide range of Galton's interests and activities. When he was in his forties, Galton turned his energies to the study of human heredity, in the process making contributions not only to genetics but also to statistical method and theory.

In his *Memories* [3], Galton describes a problem that puzzled him.

How is it possible for a population to remain alike in its features, as a whole, during many successive generations, if the ***average*** *produce of each couple resemble their parents? Their children are not alike, but vary: therefore some would be taller, some shorter than their average height; so among the issue of a gigantic couple there would be usually some children more gigantic still. Conversely as to very small couples. But from what I could thus far find, parents had issue less exceptional than themselves.**

In the 1880s data representing two generations of a human population were not available and, at the suggestion of Sir Joseph Hooker and Charles Darwin, Galton decided to experiment with sweet peas.

21.2 In 1885 Galton selected a large number of pea seeds and sorted them into seven different weight groupings. He then persuaded friends living in various parts of England to plant 70 seeds, 10 from each of the weight groups, according to a set of minute instructions. The foliage of each crop was returned to Galton, providing him with detailed data on two generations of sweet peas. The details of the experiment were presented in 1877 [2]. Forrest gives the essential results of this experiment in Table 21.1.

* Reprinted by permission of E. P. Dutton and the publishers, Methuen & Co., from *Memories of My Life* by Francis Galton.

Table 21.1 Sweet Pea Experiment

Diameter (0.01 inch) of Seed							
Parent	15	16	17	18	19	20	21
Daughter	15.4	15.7	16.0	16.3	16.6	17.0	17.3

Source. From *Francis Galton: The Life and Work of a Victorian Genius* by D. W. Forrest (Taplinger, 1974). Copyright © 1974 by Paul Elek (Scientific Books) Ltd. Reprinted by permission.

Note that the daughters of the dwarf peas are less dwarfish than the parents and the daughters of the giant peas are smaller than the parents. This phenomenon Galton called **reversion** [2]. He says that "Reversion is the tendency of the ideal mean filial type to depart from the parental type, reverting to what may roughly and perhaps fairly be described as the average ancestral type." Galton had previously used the word **regression** and had then returned to the use of the term regression.

21.3 If we express variables in standard units X' and Y' (obtained by subtracting the mean and dividing by the standard deviation), the phenomenon Galton observed can be described by Figure 21.1. Instead

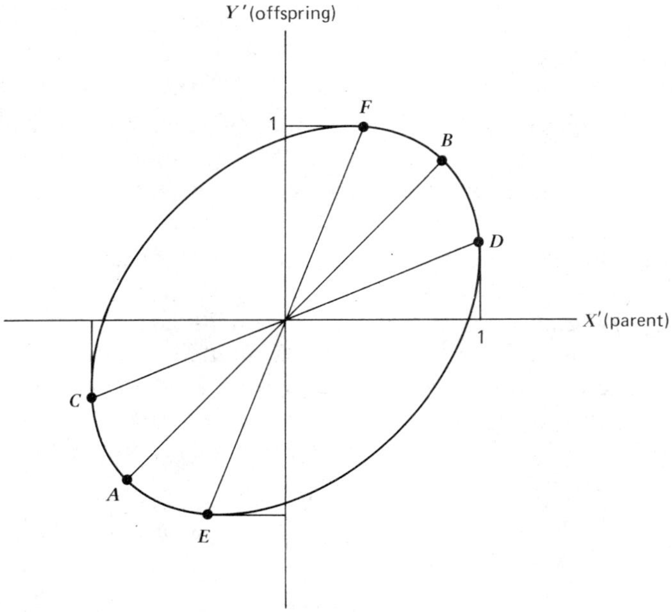

Figure 21.1 Galton's Correlation Diagram.

of data points X' and Y' being distributed about the 45-degree line AB, they tended to be distributed about the line CD. Finally, by inspection of his crude data and with mathematical assistance from friends, he came to see that a feasible explanation is given by the elliptical contours of the bivariate normal.

In Figure 21.1 we have shown a typical elliptical contour of a bivariate normal population expressed in terms of standardized variables. The major axis lies along the 45-degree line AB. Galton erected vertical lines tangent to the ellipse and connected the points C and D. Similarly, horizontal tangents produced the line EF. The logic is as follows. For any specified X', the conditional distribution of Y' is normal with the mean as the most likely value. Therefore, for a given X', the point of tangency gives the mean of the conditional distribution. Because the elliptical contours are concentric, all of the points of tangency lie on a straight line. Thus we have the remarkable fact that the **means of the conditional distributions** of Y' lie on the line CD. Similarly, the means of the conditional distribution of X' lie on the line EF.

The significance of the foregoing comments to the user of statistics is that there are **three** possible population regression lines for data arising from a bivariate normal distribution. If the $(X'Y')$ pairs were selected at random from the population we probably would be interested in the AB line. If the X' values were specified and the corresponding Y' values observed (as with the parent-offspring data) we would be interested in the CD line, called the regression line of Y on X. If, on the other hand, the values of Y' were specified and the X' values observed, we would be interested in the EF line, the regression line of X on Y.

21.4 In standardized units the equation for the mean value of Y as a function of X is simply

$$Y' = \rho X'$$

By substitution, we can obtain the equation in terms of unstandardized variables.

$$\frac{Y - \mu_y}{\sigma_y} = \rho \frac{X - \mu_x}{\sigma_x}$$

$$Y = \mu_y + \rho \frac{\sigma_y}{\sigma_x}(X - \mu_x)$$

$$= \mu_y - \rho \frac{\sigma_y}{\sigma_x}\mu_x + \rho \frac{\sigma_y}{\sigma_x} X$$

Using simpler notation, we may prefer to write

$$Y = \alpha + \beta X$$

It is helpful to visualize not only the regression but all of the conditional distributions, as in Figure 21.2. Thus for each X there is a normal population of Y values with **mean** $\alpha + \beta X$ and **variance** σ^2. The means lie on the straight line $Y = \alpha + \beta X$, and the variance is the same for every population. In fact, $\sigma^2 = \sigma_y{}^2(1 - \rho^2)$. Thus the conditional variance of Y is smaller than the unconditional variance of Y by an amount $\rho^2\sigma_y{}^2$ and by the fraction ρ^2. Considerable significance is often attached to this fact (particularly in social science statistics), and ρ^2 is said to be the fraction of variation **explained** by the regression.

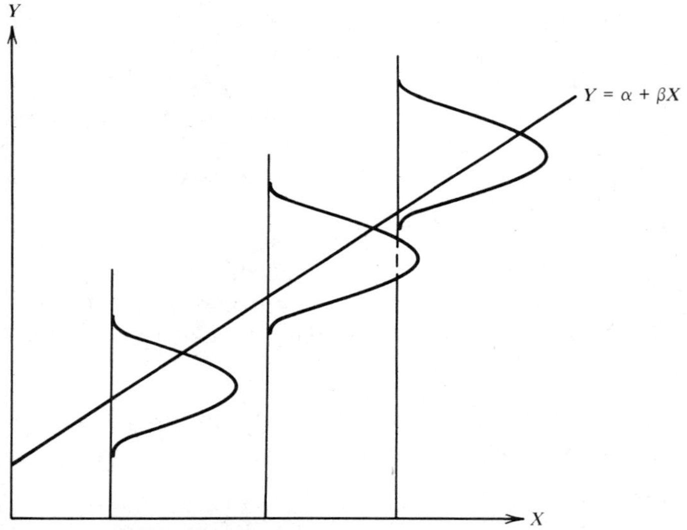

Figure 21.2 Bivariate Normal Regression.

21.5 Although the bivariate normal is of historical importance in the study of regression, it no longer plays a central role. Quite frequently it is not necessary to suppose that there is a bivariate normal population. It may be reasonable to suppose that for each X, there is a normal population of Ys, that the means lie on a straight line, and that the variance is the same for all Y populations.

SUMMARY. The puzzling association of the word **regression** with drawing straight lines through clusters of points can be explained by studying Galton's original use of the word. He observed that in many instances the characteristics of the parents tended to be passed on to the children, but that there was a **reversion** or **regression** toward the overall population mean. The word has gradually become associated with straight line data of all types.

Given data from a **bivariate normal distribution**, there are **three** regression lines of interest: the **major axis** of the ellipse, the line joining tangent points of **horizontal** tangents, and the line joining tangent points of **vertical** tangents. These lines have the following statistical interpretations: the regression line for random selection of (X, Y) pairs, the regression line for Xs selected at random for **specified** Y**s**, and the regression line for Ys selected at random for **specified** X**s**.

The last of these interpretations receives considerable emphasis and can be restated as follows. For each X, there is a population of Y values with **mean** $\alpha + \beta X$ and constant variance σ^2.

REFERENCES

1. Forrest, D. W. (1974). *Francis Galton, the Life and Work of a Victorian Genius*, New York: Taplinger.
2. Galton, Francis. (1877). "Typical Laws of Heredity," *Proceedings of the Royal Institute*, *8*, 282–301.
3. Galton, Francis. (1909). *Memories of My Life*, New York: E. P. Dutton.

LEGENDRE'S PRINCIPLE OF LEAST SQUARES

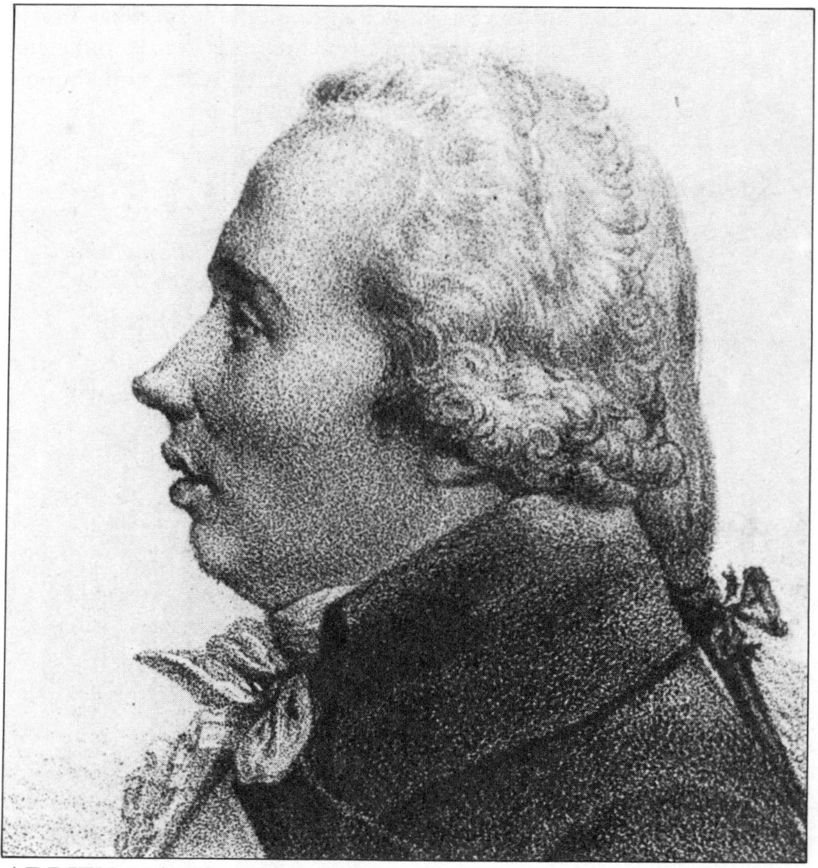

ADRIEN MARIE LEGENDRE. 1752–1833. Courtesy Columbia University Library, D. E. Smith Collection.

22

THEORY OF LEAST SQUARES

22.1 For 200 years the **method of least squares** has been known to scientists and has become the primary method by which curves and equations are fitted to data. Much research in recent years has been directed toward finding better methods of fitting, but least squares remains dominant. Perhaps this is because the idea, once acquired, seems intuitive.

In some of the simplest problems the method of least squares seems to be nothing more than choosing the most direct solution. Years ago I heard the following illustration given by Professor George Polya in a lecture at Oklahoma State University. If you airlifted a cow to the middle of a lake and deposited her there, what route would she take in swimming to shore?

I have not studied such reactions from cows, but it seems reasonable that most such "dumb" animals would swim to point A in preference to point B (Figure 22.1), for instance. Furthermore, it seems reasonable that the destination chosen on the shore would be near the point **closest** to where the swim began [map coordinates (x_1, y_1)]. In **minimizing** the distance to the shore, suppose that our cow knew algebra and that she knew the distance from (x_1, y_1) to (x_2, y_2) to be

$$D = \sqrt{(x_1 - x_2)^2 + (y_1 - y_2)^2}$$

To minimize the distance, she simply chooses x_2, y_2 to minimize D. Clearly, if D^2 is minimized, so is D, so the square root need not be considered and all that is required is to choose (x_2, y_2) so as **to minimize**

$$(x_1 - x_2)^2 + (y_1 - y_2)^2$$

It seems very natural to call this the **method of least squares** (actually the minimum of a sum of squares).

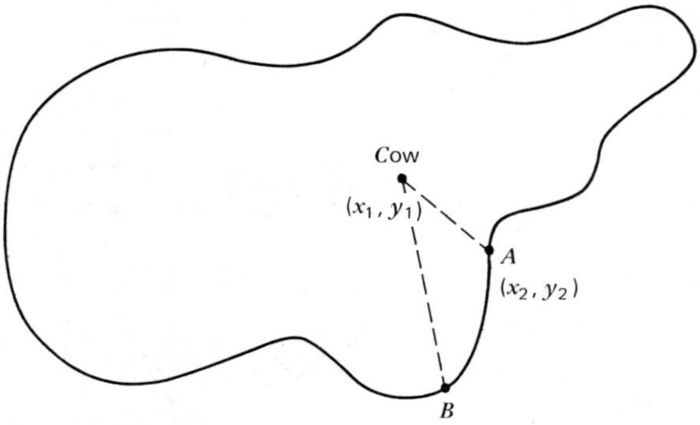

Figure 22.1 Method of Least Squares.

22.2 In order for the concept to be useful it is necessary to extend the number of dimensions. The geometric interpretation works quite nicely for three dimensions because the distance between two points (x_1, y_1, z_1) and (x_2, y_2, z_2) in three dimensions is

$$\sqrt{(x_1 - x_2)^2 + (y_1 - y_2)^2 + (z_1 - z_2)^2}$$

and least squares simply minimizes the quantity under the square root sign. For more than three dimensions we increase the number of terms under the radical, although the geometric interpretation is now purely algebraic, aided by pictures and graphs in two and three dimensions.

22.3 Gauss is widely credited for devising the method of least squares. However, it seems to have been discovered at about the same time by Gauss, Laplace, Legendre, and others. According to Plackett [2], it was first named and published by Legendre in 1806 in *Nouvelles methods pour la determination des orbites des cometes* (*new methods for the determination of comet orbits*). Abbe [1] believes that the method was invented by Gauss as early as 1795 but was first published by Legendre in 1806. Abbe goes on to argue that Robert Adrain discovered the method independently and published it in the *Analyst* in 1808 in response to the study of a prize problem in that journal.

22.4 The basic concept of least squares can (and should) be perceived apart from the methods used to achieve the minimum value. The specific minimization method used may be quite complex, making use of algorithms from nonlinear programming, or relatively simple, using first-year calculus. Simpler still, minimization may be achieved by using only algebra or by trial and error without any mathematics. It is not the purpose of this lecture to give a survey of minimization methods, but some understanding of the process is desirable. Therefore we will give some examples of minimization by elementary means.

22.5 Consider one of the simplest problems in statistics. Given n observations from a population with mean μ, find the **least squares estimate** of μ. That is, find the value $\hat{\mu}$ that minimizes

$$(x_1 - \hat{\mu})^2 + (x_2 - \hat{\mu})^2 + (x_3 - \hat{\mu})^2 + \cdots + (x_n - \hat{\mu})^2 = \sum_{i=1}^{n} (x_i - \hat{\mu})^2$$

By elementary algebra, we rewrite this quantity as follows.

$$\begin{aligned}\sum_{i=1}^{n} (x_i - \hat{\mu})^2 &= \sum_{i=1}^{n} (x_i - \bar{x} + \bar{x} - \hat{\mu})^2 \\ &= \sum_{i=1}^{n} (x_i - \bar{x})^2 + 2(\bar{x} - \hat{\mu}) \sum_{i=1}^{n} (x_i - \bar{x}) + n(\bar{x} - \hat{\mu})^2\end{aligned}$$

Now, because $\sum_{i=1}^{n} (x_i - \bar{x}) = 0$, we have

$$\sum_{i=1}^{n} (x_i - \hat{\mu})^2 = \sum_{i=1}^{n} (x_i - \bar{x})^2 + n(\bar{x} - \hat{\mu})^2$$

On the right side of the equation the first term does not involve $\hat{\mu}$, and the second term is clearly smallest when $\hat{\mu} = \bar{x}$. Therefore the least squares estimate of μ is given by $\hat{\mu} = \bar{x}$. The method of least squares thus gives us the intuitively desirable prescription of estimating the **population mean** by the **sample mean**. Furthermore, this result is obtained by elementary algebra.

22.6 Let us consider another example of minimizing a sum of squares by elementary means. In yacht racing, the effective time (used for deciding the finish) is obtained by multiplying the elapsed time by a rating factor for the boat. If we let

$$Y = \text{effective time}$$

and

$$X = \text{elapsed time}$$

then

$$Y = \beta X$$

Let us suppose that a particular boat has raced five times and that its elapsed times as well as effective times are known. They are as follows.

X	Y
2.1	2.3
2.2	2.5
3.4	3.6
6.4	7.1
5.2	5.8

We want to use the data to get a least squares estimate of β. That is, we want to minimize

$$S = \sum_{i=1}^{5} (Y_i - \beta X_i)^2$$

with respect to β, the rating factor of the boat. One simple way to proceed is by trial and error.

Inspection of the data suggests that β is about 1.1. We can see from Table 22.1 that our initial trial of $\beta = 1.1$ is very close to the value that would minimize S. Because three of the βX values are too small and two are too large, we might try a slightly larger value for β. In this way we can, by trial and error, find the least squares estimate to any number of decimal places required.

Table 22.1 Trial Calculations: $\beta = 1.1$

X	Y	$1.1X$	$Y - 1.1X$	$(Y - 1.1X)^2$
2.1	2.3	2.31	−0.01	0.0001
2.2	2.6	2.42	0.08	0.0064
3.4	3.6	3.74	−0.14	0.0196
6.4	7.1	7.04	0.06	0.0036
5.2	5.8	5.72	0.08	0.0064
Total				0.0361

22.7 Although calculus can be used to obtain a least squares formula in the example just considered, it is not necessary. We can, in fact, obtain it by elementary algebra as follows.

$$S = \Sigma(Y_i - \beta X_i)^2 = \Sigma(Y_i^2 - 2\beta X_i Y_i + \beta^2 X_i^2)$$
$$= \Sigma Y_i^2 - 2\beta \Sigma X_i Y_i + \beta^2 \Sigma X_i^2$$

Let us now **complete the square** in β, treating β as the variable and the quantities in X and Y as coefficients.

$$S = \Sigma X_i^2 \left[\beta^2 - 2\beta \frac{\Sigma X_i Y_i}{\Sigma X_i^2} + \frac{(\Sigma X_i Y_i)^2}{(\Sigma X_i^2)^2} \right] + \Sigma Y_i^2 - \frac{(\Sigma X_i Y_i)^2}{\Sigma X_i^2}$$
$$= \Sigma X_i^2 \left(\beta - \frac{\Sigma X_i Y_i}{\Sigma X_i^2} \right)^2 + \Sigma Y_i^2 - \frac{(\Sigma X_i Y_i)^2}{\Sigma X_i^2}$$

It is clear that the first term in the last expression, and therefore S, is minimized by taking $\beta = \Sigma X_i Y_i / \Sigma X_i^2$ and that the minimum value is

$$\Sigma Y_i^2 - \frac{(\Sigma X_i Y_i)^2}{\Sigma X_i^2}$$

Therefore the least squares estimate is given by

$$\hat{\beta} = \frac{\Sigma X_i Y_i}{\Sigma X_i^2}$$

In Table 22.2 the calculations for obtaining $\hat{\beta}$ are given. We can see that our trial and error result was, in fact, quite close.

Table 22.2 Least Squares Calculation

X	Y	Y^2	XY
2.1	2.3	4.41	4.83
2.2	2.5	4.84	5.50
3.4	3.6	11.56	12.24
6.4	7.1	40.96	45.44
5.2	5.8	27.04	30.16
Total		88.81	98.17

$$\hat{\beta} = \frac{99.17}{88.81} = 1.1054$$

22.8 In summary, the method of least squares minimizes an appropriate sum of squares. Many methods can be used to achieve the minimization.

SUMMARY. In the late eighteenth and early nineteenth centuries the method of **least squares** was popularized by **Gauss, Legendre, Laplace,** and others. Despite efforts to find better methods, it still remains as a fundamental method of data analysis.

The intutive idea is that of the **shortest distance** from a fixed point to a region. Because the distance formula from analytic geometry takes the square root of a sum of squares, the shortest distance is found by **minimizing a sum of squares.** Quite naturally, this method is called the method of least squares.

The least squares estimate of μ is given by minimizing $\Sigma(x - \hat{\mu})^2$ with respect fo $\hat{\mu}$. By several different methods, it can be seen that this is achieved by $\hat{\mu} = \bar{x}$. Thus the **sample mean** is the **least squares estimate** of the **population mean**.

Given the simple regression model $Y = \beta X$, the least squares estimate of β requires minimizing $\Sigma(Y - \hat{\beta}X)^2$. This gives $\hat{\beta} = \Sigma XY/\Sigma X^2$. Usually X is plotted on the horizontal axis and Y on the vertical axis so that the sum of squares of **vertical deviations** is minimized.

REFERENCES

1. Abbe, Cleveland. (1871). "Historical Note on the Method of Least Squares," *American Journal of Science and Arts, 1*, 411–415.
2. Plackett, R. L. (1972). "Studies in the History of Probability and Statistics XXIX. The Discovery of the Method of Least Squares," *Biometrika, 59* (2), 239–251.

EXERCISES

1. A good understanding of least squares can be developed by empirical studies. Consider the problem in Section 22.5. Given the data, 8, 10, 12, 9, 21, evaluate $S = \Sigma(x_i - \mu)^2$ for $\mu = 10, 11, 12, 13, 14$. Graph values of S obtained versus μ. What value of μ would appear to minimize S? What value is indicated by Section 22.5?
2. Evaluate S of Section 22.6 for values of $\beta = 1.00, 1.15, 1.10, 1.15, 1.25$. Graph S versus β. What value of β seems to minimize S?

3. Plot the following data on engineering graph paper, X on the horizontal axis, Y on the vertical.

X	Y
10	7
9	9
8	10
10	9
6	5
5	6
5	4
3	3
3	4
1	4

Using a straight edge, draw a line that looks reasonable through the data. Measure the vertical deviations from the line you have drawn and calculate the sum of squares of deviations. Repeat the process until you feel confident that your line is close to the line that minimizes the sum of squares of vertical deviations.

4. Make a three-dimensional plot of the following data, using the vertical axis for Y.

X_1	X_2	Y
1	2	21
1	3	25
1	4	29
2	2	24
2	3	28
2	4	32

Draw a plane that seems reasonable, measure the vertical deviations from the plane, and calculate the sum of squares of deviations. Proceeding in this way, find the plane by trial and error that minimizes the sum of squares of deviations.

CARL FRIEDRICH GAUSS. 1777–1855. The New York Public Library Picture Collection.

23

ESTIMATION OF REGRESSION PARAMETERS

23.1 In the preceding lectures we have developed the idea of a **population regression line** and have introduced the **method of least squares** as a way of **estimating** the population parameters. In this lecture we illustrate many of the working formulas with a numerical example.

23.2 I have frequently recorded the first and last examination scores for students in my classes. Often these sets of scores appear to follow the simple linear regression model. The following data are extracted from the examination scores of a large class.

First Exam (X)	Last Exam (Y)
63	68
65	75
72	70
73	76
80	81
85	78
86	89
93	84

We will assume that

$$Y = \beta_0 + \beta_1 X + e$$

and that the errors (e) are normal and independently distributed with mean zero and variance σ^2. We will now give formulas and illustrate the computations for this example for (1) estimates of β_0, β_1, and σ^2, (2) variance of $\hat{\beta}_0$ and $\hat{\beta}_1$, (3) confidence intervals for β_0 and β_1, and (4) tests of significance for β_0 and β_1.

23.3 In Table 23.1 we give the basic calculations for all that follows. Of course, many calculators are now available that will give the column totals of Table 23.1. With such a calculator there is no need to write the individual entries in the table. From the column totals, we obtain the next set of basic quantities.

$$\bar{X} = 617/8 = 77.125$$

$$\bar{Y} = 621/8 = 77.625$$

$$\begin{aligned}
\Sigma(X - \bar{X})^2 &= \Sigma X^2 - (\Sigma X)^2/n \\
&= 48{,}377 - 47{,}586.125 \\
&= 790.875 \\
\Sigma(Y - \bar{Y})^2 &= \Sigma Y^2 - (\Sigma Y)^2/n \\
&= 48{,}547 - 48{,}205.125 \\
&= 341.875 \\
\Sigma(X - \bar{X})(Y - \bar{Y}) &= \Sigma XY - (\Sigma X)(\Sigma Y)/n \\
&= 48{,}323 - 47{,}894.625 \\
&= 428.375
\end{aligned}$$

Table 23.1 Linear Regression Calculations

	X	Y	X^2	Y^2	XY
	63	68	3,869	4,624	4,284
	65	75	4,225	5,625	4,875
	72	70	5,184	4,900	5,040
	73	76	5,329	5,776	5,548
	80	81	6,400	6,561	6,480
	85	78	7,225	6,084	6,630
	86	89	7,396	7,921	7,654
	93	84	8,649	7,056	7,812
Total	617	621	48,377	48,547	48,323

23.4 The least squares estimates of β_0 and β_1 are found (using one of the minimization methods in Lecture 22) to be given by the following formulas.

$$\hat{\beta}_1 = \frac{\Sigma(X - \bar{X})(Y - \bar{Y})}{\Sigma(X - \bar{X})^2}$$

$$\hat{\beta}_0 = \bar{Y} - \hat{\beta}_1 \bar{X}$$

For our examination score example, we use the quantities obtained in Section 23.3 to obtain

$$\hat{\beta}_1 = \frac{428.375}{790.875} = 0.5416$$

$$\hat{\beta}_0 = 77.625 - (0.5416)(77.125)$$

$$= 35.8541$$

So the least squares line is given by the equation

$$Y = 35.8541 + 0.5416X$$

This equation is graphed in Figure 23.1 along with the original data.

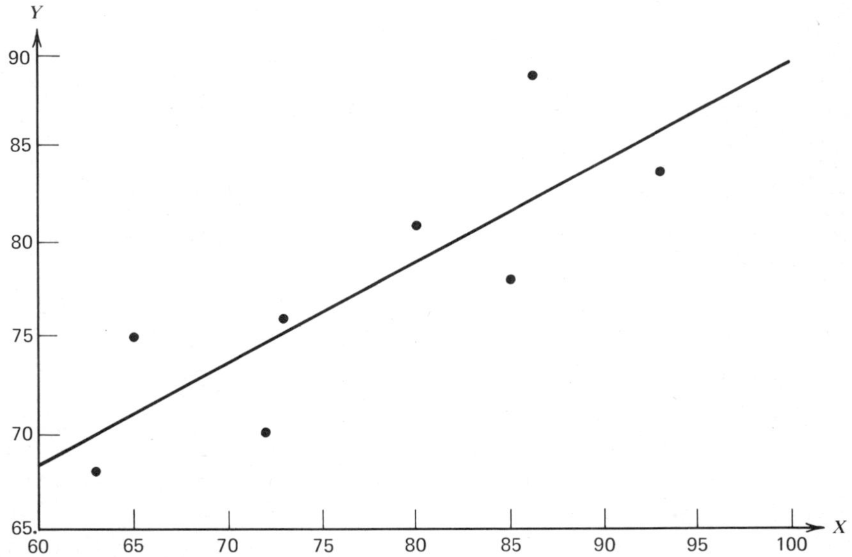

Figure 23.1 The Least Squares Line.

It may not be necessary to follow the calculations through in detail in order to understand what the line in Figure 23.1 is. Many calculators are now available, as well as computer programs, that give $\hat{\beta}_0$ and $\hat{\beta}_1$ directly. I feel that students should follow through the details of the calculations a few times in their education. The calculations should not be allowed to obscure the fundamental nature of the least squares line; i.e., it is the line for which the sum of squares of **vertical deviations** from the line is minimum.

23.5 What about these **deviations**? They are calculated by calculating the Y value on the line, denoted by $\hat{Y}$ (Y hat), and subtracting from the corresponding actual Y value. These calculations are given in Table 23.2. As a computational check, the total for the $\hat{Y}$s must equal the total for the Ys and the deviations must sum to zero. Slight discrepancies may arise because of rounding error. In the last column of Table 23.2 the deviations are squared and summed. Actually, this total is more easily obtained from the formula

$$\Sigma\, d^2 = \Sigma(Y - \bar{Y})^2 - \frac{[\Sigma(X - \bar{X})(Y - \bar{Y})]^2}{\Sigma(X - \bar{X})^2}$$

This gives

$$\Sigma\, d^2 = 109.8470$$

as obtained from the table (except for rounding error).

Table 23.2 Vertical Deviations

	X	Y	$\hat{Y} = \hat{\beta}_0 + \hat{\beta}_1 X$	$d = Y - \hat{Y}$	d^2
	63	68	69.9749	−1.9749	3.9002
	65	75	71.0581	3.9419	15.5386
	72	70	74.8493	−4.8493	23.5157
	73	76	75.3909	0.6091	0.3710
	80	81	79.1821	1.8179	3.3048
	85	78	81.8901	−3.8901	15.1329
	86	89	82.4317	6.5683	43.1426
	93	84	86.2229	−2.2229	4.9413
Total	613	621	621.0000	0.0000	109.8471

The estimate of σ^2 commonly used in statistics is

$$\hat{\sigma}^2 = \Sigma\, d^2/(n-2)$$

$$= 109.8470/6$$

$$= 18.3078$$

The divisor of $n - 2$ deserves some comment. In general, the divisor for estimating σ^2 is equal to n minus the number of parameters needed to specify the mean value. In this case β_0 and β_1 specify the mean value of Y for a particular X, so the divisor is $n - 2$.

23.6 The least squares estimates $\hat{\beta}_0$ and $\hat{\beta}_1$ are calculated from the data and would not be constant in repeated sampling from the same population. Therefore we are interested in their variances. These variances are derived in theoretical statistics and are given by

$$\sigma^2_{\beta_0} = \frac{\sigma^2 \Sigma X^2}{n\Sigma(X - \bar{X})^2}$$

$$\sigma^2_{\beta_1} = \frac{\sigma^2}{\Sigma(X - \bar{X})^2}$$

Because σ^2 is unknown, the variances of $\hat{\beta}_0$ and $\hat{\beta}_1$ are unknown. But we have an **estimate** of σ^2 and can, therefore, estimate the variances of $\hat{\beta}_0$ and $\hat{\beta}_1$. Denote these **estimated variances** by $s^2_{\hat{\beta}_0}$ and $s^2_{\hat{\beta}_1}$. These **variance estimates** are given by

$$s^2_{\hat{\beta}_0} = \frac{\hat{\sigma}^2 \Sigma X^2}{n\Sigma(X - \bar{X})^2}$$

$$= \frac{(18.3078)(48{,}377)}{(8)(790.875)}$$

$$= 139.9836$$

with the **estimated standard deviation** of $\hat{\beta}_0$,

$$s_{\hat{\beta}_0} = \sqrt{139.9836}$$

$$= 11.8315$$

and

$$s^2_{\hat{\beta}_1} = \frac{18.3078}{790.875}$$

$$= 0.0232$$

with the **estimated standard deviation** of

$$s_{\hat{\beta}_1} = 0.1523$$

23.7 The purpose of this lecture is to acquaint students with calculations, but some **interpretation** of the quantities calculated seems essential. No physical interpretation can be given to $\hat{\beta}_0$ in this example. It is simply an estimate of the Y-intercept. As far as the present example is concerned, the Y-intercept is the population mean score on the final examination of all students scoring zero on the first examination. For most classes this will not be a meaningful concept.

On the other hand, $\hat{\beta}_1$, gives an estimate of the change in final score per unit change in the first score. An increase of 10 on the first score would correspond to an increase of 5.416 on the final score.

From Figure 23.1 it is apparent that there is considerable variation of data about the least squares line. This variation results in sizable estimates of the variances of $\hat{\beta}_0$ and $\hat{\beta}_1$. With $s_{\hat{\beta}_0} = 11.8315$ we might well find 30 or 40 as acceptable values for β_0. Also, with $s_{\hat{\beta}_1} = 0.1523$ we should not attach great importance to the least squares estimate $\hat{\beta}_1 = 0.5416$ but should realize that the true value might be somewhat different.

A more formal way to approach the question of reasonable values of β_0 and β_1 is to use tests of significance. For the null hypothesis

$$H_0: \beta_0 = K$$

versus the alternative hypothesis

$$H_1: \beta_0 \neq K$$

we calculate the test statistic

$$t = \frac{\hat{\beta}_0 - K}{s_{\hat{\beta}_0}}$$

Using a t table with $6 = n - 2$ degrees of freedom, we calculate the significance level

$$\mathrm{SL} = P(|t| > |t_{\mathrm{obs}}|)$$

For null hypotheses concerning H_1, we perform analogous computations.

Suppose we have

$$H_0: \beta_1 = 0.4$$
$$H_1: \beta_1 \neq 0.4$$

Then

$$t = \frac{\hat{\beta}_1 - 0.4}{s_{\hat{\beta}_1}} = 0.9297$$

and

$$\mathrm{SL} = 0.39$$

Because this is a fairly large value, we find the hypothesized value of β_1 in good agreement with the data.

23.8 It will be apparent to the thoughtful student that $\beta_1 = 0.4$ is not the only value in agreement with the data. A recommended procedure is to calculate the significance level, SL, for various hypothesized values of β_1 and to graph SL versus hypothesized values. Values in agreement with (or in consonance with) the data can be obtained from the graph.

Most textbooks present such intervals for only one significance level. For example, if we specify a significance level of 0.05, the interval is given by

$$\hat{\beta}_1 \pm t_{0.025} s_{\hat{\beta}_1}$$

which, for the present example, is

$$0.5416 \pm (2.4469)(0.1523)$$

or

$$0.5416 \pm 0.3727$$

or

$$0.1689 \text{ to } 0.9143$$

23.9 Least squares regression lines are widely used, and observant students will find many examples. For example, Hume [2] gives the equation obtained by Binford for the **pipe stem data** of Table 5.5 to be $Y = 1931.85 - 38.26X$, where X is the hole diameter and Y is the date. This equation can be used for dating pipe stem fragments for which diameter measurements can be made.

Although the data from which the equations were obtained are not given, regression equations relating **lengths** of arm and leg bones to **height** of the person are given in the book called *The Body* [6]. These equations can be used by anthropologists for estimating what the living height of a person might have been from the measurement of one bone.

Nigel [5] made use of a least squares line in studying the behaviour of the mink in searching for food. He found a linear relationship between the **search duration** and **duration** of subsequent pursuits.

As a final example, we mention the article of Lave and Seskin [4], which discusses the use of regression in relating **air pollution** to the **human mortality rate**.

SUMMARY. The computational details of estimation, calculating confidence intervals, and performing tests are illustrated for the parameters of the **simple linear regression** model

$$Y = \beta_0 + \beta_1 X + e$$

The **least squares estimates** of β_0 and β_1 are given by

$$\hat{\beta}_0 = \bar{Y} - \hat{\beta}_1 \bar{X} \quad \text{and} \quad \hat{\beta}_1 = \Sigma(X - \bar{X})(Y - \bar{Y})/\Sigma(X - \bar{X})^2$$

The estimate of σ^2 is given by the sum of squares of vertical deviations, Σd^2, divided $n - 2$. Formulas are given for the estimated variance of $\hat{\beta}_0$ and $\hat{\beta}_1$, $s^2_{\hat{\beta}_0}$ and $s^2_{\hat{\beta}_1}$.

Confidence intervals for β_0 and β_1 are given by $\hat{\beta}_0 \pm t_{0.025} s_{\hat{\beta}_0}$ and $\hat{\beta}_1 \pm t_{0.025} s_{\hat{\beta}_1}$, respectively.

Tests of hypothesized values for the intercept and slope are given by the t test, with t values given by

$$\frac{\hat{\beta}_0 - \beta_0}{s_{\hat{\beta}_0}} \quad \text{and} \quad \frac{\hat{\beta}_1 - \beta_1}{s_{\hat{\beta}_1}}$$

respectively. In both cases the degrees of freedom are $n - 2$.

Examples are given of the use of simple linear regression in **archaeology**, **anthropology**, **animal behavior**, and **ecology**.

REFERENCES

1. Draper, N., and H. Smith. (1966). *Applied Regression Analysis*, New York: Wiley.
2. Hume, Ivor Noël. (1970). *A Guide to Colonial Artifacts*, New York: Knopf.
3. Kempthorne, O., and L. Folks. (1971). *Probability, Statistics, and Data Analysis*, Ames: Iowa State Press.
4. Lave, Lester B., and Eugene P. Seskin. (1979). "Epidemiology, Causality, and Public Policy," *American Scientist*, *67*, 178–186.
5. Nigel, Dunstone. (1978). The Fishing Strategy of the Mink," *Behavior*, *67*, 157–177.
6. Nourse, Alan E. (1964). *The Body*, New York: Time.

EXERCISES

1. The record times for the mile run from 1911 to 1975 are as follows.

1911	4:15.4	1945	4:01.4
1913	4:14.6	1954	3:59.4
1915	4:12.6	1954	3:58.0
1923	4:10.4	1957	3:57.2
1931	4:09.2	1958	3:54.5
1933	4:07.6	1962	3:54.4
1934	4:06.8	1964	3:54.1
1937	4:06.4	1965	3:53.6
1942	4:06.2	1966	3:51.3
1942	4:04.6	1967	3:51.1
1943	4:02.6	1975	3:51.0
1944	4:01.6	1975	3:49.4

a. Convert the times to seconds.
b. Plot a scattergram of the record time (Y) versus the year (X).
c. Assuming the straight line model $Y = \beta_0 + \beta_1 X + e$, estimate β_0, β_1, and σ^2.
d. Calculate the significance level for $H_0: \beta_1 = -0.4$ versus $H_1: \beta_1 \neq -0.4$.
e. Predict the record time for the mile run in the year 2000.
f. What reservations do you have about using a straight line?

2. The following figures on total federal tax revenue, in billions of dollars, are given by the *1976 Statistical Abstract of the United States.*

1950	44	1969	200
1955	72	1970	206
1960	100	1971	203
1965	126	1972	223
1968	165	1973	248

   a. Plot a scattergram of tax (Y) versus the year (X).
   b. Estimate β_0, β_1, and σ^2.
   c. Calculate the significance level for $H_0: \beta_1 = 10$ versus $H_1: \beta_1 \neq 10$.
   d. Predict the total tax for the year 2000.
   e. What reservations do you have about using a straight line?

3. Calculate the least squares line for the data of Table 5.5. Does your equation agree with the equation given by Hume?

4. For each of the students in your class, obtain X, the length of the radius (the arm bone from the elbow to the wrist) and Y, the height. Calculate the least squares regression line.

5. The total federal outlay in billions of dollars for the national defense and veterans benefits was as follows for 1965 to 1972.

Year	Outlay
1965	54
1966	62
1967	76
1968	86
1969	88
1970	88
1971	87
1972	88

   a. Plot the data.
   b. Calculate the least squares straight line.
   c. Using the least squares line, what would you predict for the outlay in 1977? The actual outlay was 119.

6. The time required to construct a house is related to the complexity and, therefore, the price of a house. The following data represent

completion times, X, in months, and the price, Y, in thousands of dollars.

X	Y
12	100
10	60
6	50
7	60
10	90
11	110
12	120
8	70
7	60

Using the least squares line, about what should a new house cost if the contractor is estimating a construction time of $10\frac{1}{2}$ months?

7. In response to a market survey several motorists report the tire pressure that they strive to maintain and the number of miles service received on the original tires. The average pressure in pounds is indicated by X and the miles (thousands) are recorded by Y.

X	Y
29	22.5
28	21.5
31	25.0
32	26.3
27	21.2
26	20.8
30	24.0

a. Calculate the least squares line.
b. Test the null hypothesis that $\beta_1 = 1.0$ versus the alternative that $\beta_1 \neq 1.0$. Use $\alpha = .05$.
c. What mileage would you expect at 29 pounds pressure?

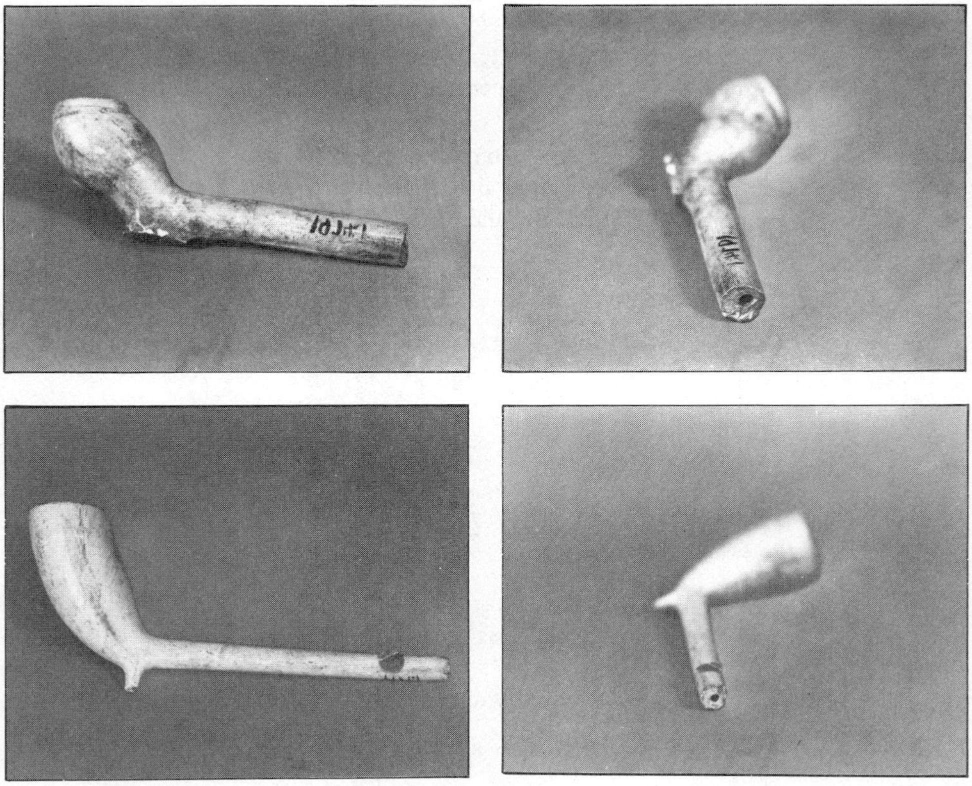

ENGLISH CLAY PIPES. Predicting the age of the pipe from the diameter of the pipe stem bore. U.S. Tobacco Museum, Greenwich, Connecticut.

24

PREDICTION

24.1 In the previous lecture we concerned ourselves primarily with the mechanics of obtaining the least squares line, although we also presented methods of forming opinions about reasonable values for the βs in light of the data.

One of the very important uses of regression lines is the **prediction** of **future** Y values. We cite as examples or possible examples (1) the prediction of college grade point average, Y, from entrance examination scores, X, (2) salesmen productivity, Y, from aptitude test scores, X, (3) weight gain of cattle, Y, from number of days on ration, X, etc. The most reasonable single predicted value of Y is simply

$$\hat{Y} = \hat{\beta}_0 + \hat{\beta}_1 X$$

We may also wish to give an interval of reasonable values.

A second problem concerns estimating the **mean** of the Y population for a specified X. Again, the single estimated value is simply

$$\hat{Y} = \hat{\beta}_0 + \hat{\beta}_1 X$$

but an interval of reasonable values is different than if we are interested in a single Y value. For the examination score example of Lecture 23 we might be interested in the final mean score of the population of students scoring 70 on the first examination. This is obviously different than being interested in the final score of a single student scoring 70 on the first examination, and the intervals are different.

A third problem is to give intervals for the mean of all Y populations **simultaneously**. This is equivalent to specifying a region in which the entire true regression line lies with reasonable assurance.

A fourth problem of importance in some areas of application is the estimation of an unknown X for a known Y value. This is the so-called problem of **calibration**, which arises in the following way. Two measurements, X and Y, are available on a number of specimens. The measurement X is quite precise, but it is difficult or expensive. The measurement Y is less precise, but it is easy or inexpensive. The least squares line $Y = \hat{\beta}_0 + \hat{\beta}_1 X$ is obtained and is used to estimate the X value from measuring Y on future specimens.

24.2 Suppose that an instructor has been teaching large lecture sections for several years and, while using standardized examinations, has obtained a regression line relating course averages (Y) to the first examination scores. A student has scored 70 on the first examination and is thinking about dropping the course. The instructor uses the regression line and the first score of the student to obtain a prediction interval for the student's course average score. The end points of the interval are given by

$$\hat{\beta}_0 + \hat{\beta}_1 X_0 \pm t_{\text{tab}}\hat{\sigma}\sqrt{1 + \frac{1}{n} + \frac{(X_0 - \bar{X})^2}{\Sigma(X - \bar{X})^2}} \qquad (1)$$

where X_0 = the X value of interest (70 in this case), t_{tab} has $n - 2$ degrees of freedom, and the other quantities are calculated from the regression data.

We carry through the calculations using the data from Lecture 23, letting Y from that data represent average course scores. Then

$$\bar{X} = 77.125 \qquad \Sigma(X - \bar{X})^2 = 790.875$$

$$n = 8 \qquad \hat{\sigma}^2 = 18.3078$$

$$\hat{\beta}_0 = 35.8541 \qquad \hat{\beta}_1 = 0.5416$$

$$\sqrt{1 + \frac{1}{n} + \frac{(X_0 - \bar{X})^2}{\Sigma(X - \bar{X})^2}} = 1.0905$$

So the end points of a 95% prediction interval are given by

$$35.8541 + (70)(0.5416) \pm (2.4469)(4.2788)(1.0905)$$

or

$$73.766 \pm 11.4173$$

or

$$62.3488 \quad \text{and} \quad 85.1834$$

Thus the 95% prediction interval for the student in question is from 62.3488 to 85.1834. In a real situation such a prediction interval could be of value in counseling a student. The student might well question the assurance with which we predict the final score to lie between the specified limits. We could tell him or her that (given the assumptions), 95% of all such prediction intervals for students with an initial score of 70 would contain their course average.

24.3 Suppose that we wished not to predict for a single student but to estimate the mean course score of all students with an initial score of 70. It seems quite intuitive that we should be able to locate the mean score more precisely than that of a single student. In fact, the confidence interval for the Y population mean at a given X value is

$$\hat{\beta}_0 + \hat{\beta}_1 X_0 \pm t_{\text{tab}} \hat{\sigma} \sqrt{\frac{1}{n} + \frac{(X_0 - \bar{X})^2}{\Sigma(X - \bar{X})^2}} \qquad (2)$$

It is immediately apparent that this formula is almost the same as that in the previous section. Closer inspection reveals that the "1" under the radical sign in the previous equation has been removed. So the quantity under the radical sign is smaller, and we are adding and substracting a smaller quantity to $\hat{Y}$ to get the end points of the confidence intervals. Carrying the calculations through for 95% confidence, we find

$$\sqrt{\frac{1}{n} + \frac{(X_0 - X)}{\Sigma(X - \bar{X})^2}} = 0.4350$$

so the end points of our interval are

$$73.7661 \pm (2.4469)(4.2788)(0.4350)$$

or

$$73.7661 \pm 4.5544$$

or

$$69.2117 \quad \text{and} \quad 78.3205$$

In the context of our example we would be 95% confident that the course average for the population of students scoring 70 on the first examination lies between 69.2117 and 78.3205. So, in fact, the interval for the Y population mean is considerably shorter than the interval for a single predicted observation.

24.4 Both of the formulas given by Equations 1 and 2 are intended to be used with a **single** X value. If used with two or more values the confidence to be attached to 95% intervals is **not** 95% for all of the intervals jointly. However, a formula does exist for setting joint confidence intervals on the mean of the Y populations for all X, i.e. on $\beta_0 + \beta_1 X_0$ for all X_0. The formula is

$$\hat{\beta}_0 + \hat{\beta}_1 X_0 \pm \sqrt{2F_{\text{tab}}}\,\hat{\sigma}\sqrt{\frac{1}{n} + \frac{(X_0 - \bar{X})}{\Sigma(X - \bar{X})^2}} \qquad (3)$$

where F_{tab} is the tabulated F value for the desired level of confidence with 2 degrees of freedom for the numerator and $n - 2$ degrees of freedom for the denominator.

Because this is our first mention of F, some further comments seem in order. The F distribution is concentrated on the nonnegative numbers and is skewed to the right. Although the derivation is due to Fisher [2], one of the first tables was due to Snedecor [3], and the distribution is widely known as Snedecor's F. The distribution is tabled in Table A.7, where v_1 = numerator degrees of freedom and v_2 = denominator degrees of freedom.

It should be apparent that a wider interval is required if we wish to set confidence limits simultaneously on all values of $\beta_0 + \beta_1 X$ than if we are satisfied with one X only. Comparison of Equations 2 and 3 shows that this does happen. In Equation 3 we multiply by $\sqrt{2F_{\text{tab}}}$ instead of t_{tab}. Because $\sqrt{F_{\text{tab}}}$ is about the same as t_{tab}, the major difference between 2 and 3 is determined by $\sqrt{2}$. Thus Equation 3 gives an interval at least 40% wider than that given by Equation 2.

Carrying through the arithmetic for $X_0 = 70$ with the current example, we find $\sqrt{2F_{\text{tab}}} = 3.2073$ for 95% confidence, and the end points of the interval are given by

$$73.7661 \pm (3.2073)(4.2788)(0.4350)$$

or

$$73.7661 \pm 5.9697$$

or

$$67.7964 \quad \text{and} \quad 79.7358$$

SUMMARY. Implicitly or explicitly, formally or informally, regression lines are used to **predict** the relationship between X and Y, other than those observed. Confidence intervals are calculated for (1) a single future Y value for a specified X, (2) the Y population mean for a specified X, and (3) the Y population means for all X simultaneously (the entire population regression line).

The three formulas are as follows.

$$\hat{Y} \pm t\hat{\sigma}\sqrt{1 + \frac{1}{n} + \frac{(X_0 - \bar{X})^2}{\Sigma(X - \bar{X})^2}} \tag{1}$$

$$\hat{Y} \pm t\hat{\sigma}\sqrt{\frac{1}{n} + \frac{(X_0 - \bar{X})^2}{\Sigma(X - \bar{X})^2}} \tag{2}$$

$$\hat{Y} \pm \sqrt{2F}\,\hat{\sigma}\sqrt{\frac{1}{n} + \frac{(X_0 + \bar{X})^2}{\Sigma(X - \bar{X})^2}} \tag{3}$$

Equation 1 for individual Y values gives a wider interval than Equation 2 for a mean Y value. Equation 3 for the entire regression line gives wider intervals than Equation 2 for a single mean Y value.

REFERENCES

1. Draper, N. R., and H. Smith. (1966). *Applied Regression Analysis*, New York: Wiley.
2. Fisher, R. A. (1933). *Statistical Methods for Research Workers*, Edinburgh: Oliver & Boyd.
3. Snedecor, George W. (1934). *Calculation and Interpretation of Analysis of Variance and Covariance*, Ames, Ia.: Collegiate Press.

EXERCISES

1. Using the data from Exercise 1 of Lecture 23, construct a 95% confidence interval for the mile run record time for the year 1990 (use Equation 1). Construct a 95% confidence interval for the record time in 1990 using Equations 2 and 3 and give interpretations.
2. Using the data from Exercise 2 of Lecture 23, construct a 95% confidence interval for the total federal tax revenue in the year 2000.
3. Using the data of Table 5.5, construct a 95% confidence interval for the date of a pipe for which the pipe stem measurement is $\frac{6}{64}$ inch.

4. Using the data of Exercise 4 of Lecture 23, construct a 95% confidence interval for the height of a person with a radius measurement of 11 inches and for the average height of all people with a radius measurement of 11 inches.
5. An office supervisor monitors carefully the work of typists under different working conditions. The average number of errors per page during the day and the average office temperature that day are recorded for several different typists. The data are:

Temperature (X)	Errors (Y)
68.2	4.0
69.1	3.9
71.2	3.4
64.3	5.2
68.2	4.1
69.3	4.8
65.0	6.1
68.2	4.4
70.3	3.6
66.7	6.2
68.3	4.2
69.2	4.6

Calculate a 95% confidence interval for the average number of errors per page for all typists and for a single typist in an office kept at 65°F.
6. In a large department store storewide sale, the following prices (dollars), regular (X) and sales (Y), were observed on a sample of items:

X	Y
15.00	13.50
12.50	11.00
56.00	35.00
75.00	42.00
89.00	47.50
77.00	42.00
95.00	57.50
25.00	20.00
37.50	25.00
49.50	29.50

Calculate a 95% confidence interval for the sales price of a single item regularly priced at \$44.50.

7. A monthly magazine reports the average monthly temperature (°F) and the average relative humidity (%) for several American cities. Calculate the 95% confidence interval for the average relative humidity for a single city with an average monthly temperature of 60°F where X = temperature, and Y = relative humidity.

X	Y	X	Y
35	57	68	74
38	63	66	78
43	59	59	79
40	68	58	72
45	67	54	66
46	74	53	74
50	61	49	69

STATISTICS OF EXPERIMENTAL SCIENCE

ROTHAMSTED MANOR HOUSE. Ancestral home of Sir John Bennet Lawes, founder of Rothamsted Experimental Station. Reproduced from a print from the Royal Commission on Historical Monuments, England. Reprinted by permission from the librarian of Rothamsted Experiment Station.

25

THE LEGACY OF ROTHAMSTED AND AMES

25.1 Sir John Bennet Lawes (1814–1900) was born in the Rothamsted Manor House near Harpenden, England [2]. After studying at Oxford, he returned to the manor, which had been in his family for 200 years, to take up the management of the family farm in 1834. Because of his interest in **chemistry**, he shortly began to conduct experiments on the effect of chemical salts on **plant growth**. He received a patent for the manufacture of superphosphate in 1842 and opened his first factory for its manufacture at Deptford Creek, London, in 1843. In the same year he appointed a chemist, J. H. Gilbert, as his assistant, and the systematic study of **crop nutrition** at Rothamsted was begun. The experiments on wheat begun in 1843 on Broadbalk field (which continue to this day) were the first of many experiments involving most crops grown in England.

In 1889 Lawes set up the Lawes Agricultural Trust to insure the continuation of the Rothamsted experiments. He endowed the trust with a sum of 100,000 pounds and granted a 100-year lease to the fields on which the classical experiments were conducted.

The contribution of Rothamsted to agricultural science is apparent. But why should the casual students of statistics be aware of its existence? Because in 1919 there occurred an event of singular importance in the history of statistics—**R. A. Fisher** came to Rothamsted.

25.2 The story of how Fisher came to Rothamsted is told by Sir E. John Russell [7], who was the Director at the time. He writes:

But Cambridge had nothing to offer him and he left in 1913. For the next two years he was unsettled; he volunteered for military service in the 1914–18 war but was rejected on account of his eyesight, and for some time he *worked on a farm in Canada. From 1915 to 1919 he taught mathematics and physics at various public schools; he was constitutionally unfitted for the work and was neither happy or successful at it: when I first saw him in 1919 he was out of a job. Before deciding anything I wrote to his tutor at Caius College whom I knew personally, asking him about his mathematical ability. The answer was that he could have become a first class mathematician had he "stuck to the ropes," but he would not. That looked like the type of man we wanted, so I invited him to join us. I had only £200, and suggested that he should stay as long as he thought that the sum should suffice, and after studying our records he should tell me whether they were suitable for proper statistical examination and might be expected to yield some more information than we had extracted. He reported weekly at tea at my house and always favourably. It took me a very short time to realize that he was more than a man of great ability: he was in fact a genius who must be retained.***

25.3 Fisher was at Rothamsted from 1919 to 1933. During those 14 years he made almost immeasurable contributions to the rapidly developing subject of statistics. His papers on theoretical statistics still form the foundation of much of modern statistics. Statistical methods devised by Fisher are known and used throughout the world. His book, *Statistical Methods for Research Workers* [3], first published in 1925, has gone through 14 editions; the last one was published in 1970, after Fisher's death in 1965. Some measure of the tremendous outpouring of Fisher's genius is given by the *Collected Papers of R. A. Fisher* [1], a five-volume set containing over 300 separate articles. It is difficult to comprehend the full sweep of Fisher's work but, in the next several lectures many of the ideas of **experimental design** that we present can be traced either to Fisher or his associates and followers. Perhaps the greatest legacy to statistics (or science in general) from Fisher's days at Rothamsted is Fisher's work on experimental design as exemplified by his paper, "The Arrangement of Field Experiments" [4], and his book, *The Design of Experiments* [5], first published in 1935.

25.4 On the other side of the Atlantic in Ames, Iowa, a young associate professor of mathematics, George Waddell Snedecor, had become interested in statistics. Snedecor was born in Memphis, Tennessee, in 1881 but spent most of his childhood and early adulthood in Florida and

* Reproduced from *The History of Agricultural Science in Great Britain*, by Sir E. John Russell. Copyright © 1966 by George Allen & Unwin (Publishers) Ltd. Reprinted by permission from George Allen & Unwin (Publishers) Ltd.

Alabama. He received a bachelor's degree in mathematics and physics from the University of Alabama in 1905 and a master's degree in physics from the University of Michigan in 1913. In 1913 he was appointed assistant professor of mathematics at Iowa State University in Ames: he was made associate professor the following year.

As the interest in statistics at Ames developed, he was given responsibility for teaching statistics courses; the first statistics course in the mathematics department was offered in 1914 to 1915.

In the early 1920s Snedecor began a collaboration with Henry A. Wallace, who was later to become vice-president of the United States. Wallace was editor of *Wallace's Farmer* in Des Moines, Iowa, at the time and was interested in agricultural research. In the spring of 1924 Snedecor organized a Saturday afternoon seminar to study **multiple regression** and other statistical methods under the guidance of Wallace. From these Saturday afternoon seminars came the publication in 1925 of *Correlation and Machine Calculation* by Wallace and Snedecor.

Snedecor and his colleagues were pioneering not only in statistics but in the use of punch cards and punch card machines to analyze data. This effort led to establishing the Mathematical Statistical Service in 1927 and the Statistical Laboratory in 1933.

In 1931 the first Master of Science degree in statistics was awarded to Snedecor's student Gertrude Cox, who would go on to establish the Institute of Statistics in North Carolina. In 1931 Snedecor invited Fisher to Ames to teach during the summer session. This summer session affected in many ways the development of statistical methods at Ames and around the world.

In 1934 Snedecor published a book called *Calculation and Interpretation of Analysis of Variance and Covariance.* In 1937 there followed the first edition of his now classic *Statistical Methods.* This book has now gone through six editions, the sixth coauthored with William G. Cochran, and in 1974 had sold over 127,000 copies.

Snedecor's contributions were great. The statistics program at Ames attracted outstanding statisticians from all over the world, and the Department of Statistics at Iowa State University was the first in the United States.

SUMMARY. The **Rothamsted Experimental Station** was established by **Sir John Bennet Lawes** on his family estate near Harpenden, England. The first experiments on wheat were begun in 1843. In 1889 Lawes established a trust fund to ensure the continuation of the experiments.

When **Fisher** went to work at Rothamsted in 1919, the experiment station was already world famous. Because of Fisher's work there from 1919 to 1933, this fame embraced statistics as well as agricultural research. At Rothamsted, Fisher developed basic **statistical methods** as well as the **theoretical** foundations.

Snedecor, at Iowa State College, became interested in statistics in the early 1900s. This interest led to the organization of a consulting service in 1927 and the **Statistical Laboratory** in 1933. Because of Snedecor's interest in statistics, he invited Fisher to the United States in 1931. This visit had a great effect on statistics at Iowa State University and in the United States.

REFERENCES

1. Bennett, J. H. (1972). *Collected Papers of R. A. Fisher*, Adelaide, Australia: The University of Adelaide.
2. Boalch, D. H. (1953). *The Manor of Rothamsted*, Cambridge, England: Cambridge University Press.
3. Fisher, R. A. (1925). *Statistical Methods for Research Workers*, 1st Edition, Edinburgh: Oliver & Boyd.
4. Fisher, R. A. (1926). "The Arrangement of Field Experiments," *Journal of the Ministry of Agriculture, England, 33*, 503–513.
5. Fisher, R. A. (1935). *The Design of Experiments*, Edinburgh: Oliver & Boyd.
6. Kempthorne, O. (1974). George W. Snedecor, *International Statistical Review, 42*, 319–321.
7. Russell, Sir E. John. (1966). *The History of Agricultural Science in Great Britain*, London: Allen & Unwin.
8. Snedecor, George W. (1937). *Statistical Methods*, 1st Edition, Ames: Iowa State Press.

RANDOMIZATION. Games of chance that led to probability theory are used in experimental design. Chuck Fishman/Leo deWys, Inc.

26

RANDOMIZATION

26.1 Suppose one finds a magnetic compass that has been exposed to the weather. The needle no longer turns because of rust and corrosion. Is it reasonable to believe that the needle is pointing toward the north? If one were hopelessly lost in a wood one might be inclined to act on the hope that the needle is pointing north. In fact, without knowledge of how the compass came to be in its present state, it is impossible to have a strong conviction about what direction the needle is pointing. We would have a totally different opinion if we were using a tried and proven compass on which the needle had come to rest after being spun gently.

An analogous situation sometimes occurs where data seem to be pointing toward some particular conclusion. Fisher perceived that **randomization** provided a way of collecting data that would allow forceful conclusions—a way of "spinning the needle."

26.2 Let us then suppose, instead of finding a compass, that we come across an experiment on the effect of two different fertilizers on the yield of wheat. Three experimental plots have been given treatment 1, T_1, and three plots treatment 2, T_2. The yield (in bushels per acre) has been recorded as follows.

T_1	T_2
36.4	38.7
40.2	39.1
36.2	36.6

Beyond reporting that there is a difference in average yield of 0.53 bushels, how is one to proceed? Do the data clearly point to the superiority of treatment 2? Since we do not know how the data were arrived at, we cannot be sure. Perhaps treatment 2 was assigned to the more fertile plots of ground or perhaps the treatment 1 plots were next to a competing crop.

One way of proceeding with forming an opinion is to consider what might happen if the treatments were **assigned at random** to the experimental plots. Suppose that this were done and, in fact, treatments 1 and 2 were equal in their effect on the yield of wheat. Then the difference between yields obtained represents properties of the plots only, and the labeling by T_1 and T_2 is purely a **random labeling**. Let us consider the 20 different assignments of T_1 and T_2 that could be made to these six plots. In Table 26.1 these assignments are shown along with the resulting means $\bar{T}_1$ and $\bar{T}_2$ and the differences $\bar{T}_2 - \bar{T}_1$.

The difference actually observed in average yields, 0.53 bushels per acre, is now more meaningful when we consider the column of 20

Table 26.1 Randomization of Treatments[a]

Outcome	36.4	40.2	36.2	38.7	39.1	36.6	$\bar{T}_1$	$\bar{T}_2$	$\bar{T}_2 - \bar{T}_1$
1	T_1	T_1	T_1	T_2	T_2	T_2	37.60	38.13	0.53
2	T_1	T_1		T_1			38.43	37.30	−1.13
3	T_1	T_1			T_1		38.57	37.17	−1.40
4	T_1	T_1				T_1	37.73	38.00	0.27
5	T_1		T_1	T_1			37.10	38.63	1.53
6	T_1		T_1		T_1		37.23	38.50	1.27
7	T_1		T_1			T_1	36.40	39.33	2.93
8	T_1			T_1	T_1		38.07	37.67	−0.40
9	T_1			T_1		T_1	37.23	38.50	1.27
10	T_1				T_1	T_1	37.37	38.37	1.00
11		T_1	T_1	T_1			38.37	37.37	−1.00
12		T_1	T_1		T_1		38.50	37.23	−1.27
13		T_1	T_1			T_1	37.67	38.07	0.40
14		T_1		T_1	T_1		39.33	36.40	−2.93
15		T_1		T_1		T_1	38.50	37.23	−1.27
16		T_1			T_1	T_1	38.63	37.10	−1.53
17			T_1	T_1	T_1		38.00	37.73	−0.27
18			T_1	T_1		T_1	37.17	38.57	1.40
19			T_1		T_1	T_1	37.30	38.43	1.13
20				T_1	T_1	T_1	38.13	37.60	−0.53

[a] T_2 is assigned to the blank spaces.

possible differences that could have been observed. First, note occurrence of the pairs of differences: 0.53 and −0.53, −1.13 and 1.13, −1.40 and 1.40, etc. This is entirely reasonable because if the treatments were equal in effect, the sign of the difference is purely due to the labeling, a random effect. Second, note that 18 of the 20 differences are as large or larger in absolute magnitude than the observed difference. That is,

$$P\,(\text{absolute difference} > 0.53 \mid \text{equal effects}) = 0.90$$

Because of the high probability, the idea or hypothesis of equal treatment effect seems quite plausible.

26.3 We have just performed what was previously called a **test of significance** in Lecture 17. The general idea is that one devises a test statistic (the absolute difference of means in this example) and calculates the probability of statistic values as unlikely as that observed, given that the null hypothesis is true. This probability, called the significance level, provides a **credibility index** for the null hypothesis. Large values support and small values tend to negate the null hypothesis.

It should be emphasized that the analysis performed in the previous section could be performed **regardless** of how the treatments were assigned to the experimental plots. However, Fisher recognized that the act of **actually** assigning the treatments at random makes the analysis far more meaningful. How it does this is not clear. Perhaps it is psychological; it is not purely mathematical. We would like to say that it is analogous to the compass. Experimental results that point in a particular direction are more readily understood if we know that equal opportunity has been provided for **all** directions.

26.4 In addition to **increasing the validity** of the test of significance already described, randomization provides a **probability framework** that enables us to obtain unbiased estimates of treatment differences and to estimate unbiasedly the variance of our estimate of treatment differences. To understand this, let us suppose that the basal yields (yields that would be obtained in a given year without fertilizer) for our six plots are, in fact, the yields used in Table 26.1. Let us also suppose that the **effect** of treatment 1, τ_1, is to add 1.0 bushel per acre, and the **effect** of treatment 2, τ_2, is to add 3.0 bushels per acre. Then the 20 possible values for $\bar{T}_2 - \bar{T}_1$ would be those in Table 26.1 plus $\tau_2 - \tau_1 = 3.0 - 2.0 = 1.0$. That is, the 20 possible differences would be 1.53, −0.13, −0.40, 1.27, 2.53, 2.27, 3.93,

0.60, 2.27, 2.00, 0.00, −0.27, 1.40, −1.93, −0.27, −0.53, 0.73, 2.40, 2.13, and 0.47. The average of all possible values for $\bar{T}_2 - \bar{T}_1$ is then $\tau_2 - \tau_1 = 1.0$ bushel per acre. What has randomization achieved for us? It gives us a clearly defined population of numbers (the possible values for $\bar{T}_2 - \bar{T}_1$) for which the mean value is $\tau_2 - \tau_1$, the **parameter** of interest. Furthermore, we have, in effect, selected one of the possible values for $\bar{T}_2 - \bar{T}_1$ at random. In repetitions of the experiment we would obtain, on the average, a difference of $\tau_2 - \tau_1$. Thus randomization assures us of an unbiased estimate of the treatment difference $\tau_2 - \tau_1$.

The variance of the population of 20 differences is calculated by the formula $\Sigma(D_i - \bar{D})^2/20$, where D_i = a typical difference. Thus we find $\mathrm{Var}(\bar{T}_2 - \bar{T}_1) = 1.89$. Just as randomization makes possible an unbiased estimate of $\tau_2 - \tau_1$, it also provides us with an unbiased estimate of $\mathrm{Var}(\bar{T}_2 - \bar{T}_1)$. Suppose that the randomization were that of the first line in Table 26.1. Then the results would be

T_1	T_2
38.4, 42.2, 38.2	41.7, 42.1, 39.6

First, we estimate the inherent variability **within** the observations from T_1 and within those from T_2. Then we **pool** our estimates.

$$\text{From } T_1,\ {s_1}^2 = 5.08$$

$$\text{From } T_2,\ {s_2}^2 = 1.80$$

$$\text{Average } s^2 = 3.44$$

Our estimate of $\mathrm{Var}(\bar{T}_2 - \bar{T}_1)$ is then given by

$$\tfrac{2}{3}s^2 = 2.29$$

26.5 In summary, randomization **increases the validity** of a test of significance, **provides a well-defined population** with a mean of $\tau_2 - \tau_1$ and a variance of $\mathrm{Var}(\bar{T}_2 - \bar{T}_1)$, and **allows unbiased estimates** of the treatment difference and of the variance of $(\bar{T}_2 - \bar{T}_1)$.

We have already mentioned that Fisher is credited with the use of randomization in experimental design. However, the idea of **random sampling** from a population was known and used before Fisher. For example, Peirce [1] describes induction as "reasoning from a sample taken at random to the whole lot sampled." He insists on two rules of inductive inference: (1) the sample must be random, and (2) the proposed investigation of the sample must not be suggested by the characteristics of the sample.

SUMMARY. The physical act of **randomly** assigning treatments to experimental plots justifies the use of a **probability model** for the observations.

If the treatments have **equal effects**, the labeling of observations by treatment received can be viewed strictly as a **random labeling**. The difference between sample treatment means may be due purely to the particular labeling chosen. Therefore it is reasonable to consider all of the **possible treatment assignments** and the different values of the sample treatment means that might result from a **given** set of observations. This leads to an **observed significance level**: the probability of a treatment difference as large as that observed.

If treatments have different effects, consideration of all possible random treatment assignments shows that the sample treatment differences estimate **unbiasedly** the population treatment differences. The population of sample treatment differences also has a variance that can be estimated unbiasedly from a sample.

REFERENCE

1. Peirce, C. S. (1957). *Essays in the Philosophy of Science*, edited by Vincent Tomas, New York: The Liberal Arts Press.

EXERCISES

1. Verify the values for $\bar{T}_2 - \bar{T}_1$ in Table 26.1.
2. Show that when $\tau_1 = 2.0$ and $\tau_2 = 3.0$, as in Section 26.4, the average of all possible values for $\bar{T}_2 - \bar{T}_1$ is 1.0.
3. In Section 26.4 verify that $s_1^2 = 5.08$ and $s_2^2 = 1.80$ and, therefore, that $\mathrm{Var}(\bar{T}_2 - \bar{T}_1)$ is estimated by 2.29.

4. Design and conduct a simple experiment involving two treatments assigned at random. Estimate the treatment difference and then estimate the variance of your estimate. For example, you might study the effects of additives on automobile mileage, weight losses from different diets, or cleansing power of different detergents.
5. A car club designs a small experiment to compare the mileage obtained from two different blends of gasoline. Blend A is assigned at random to two of five cars and blend B to the other three cars. The average mileage obtained during the experiment is recorded for each of the five cars.

Blend A	Blend B
22.3	24.2
20.7	25.1
	22.6

Calculate the observed significance level for the hypothesis that there is no difference between the average mileage for the two blends of gasoline. Follow the procedure in Section 26.2.
6. In a course with multiple sections, three sections are selected at random. In these sections the instructor gives a quiz at the end of each hour. In the other three sections no quizzes are given. At the end of the course the students in all six sections take a common examination. The average final test scores are given for the two methods of instruction.

Quiz	No Quiz
82.1	78.9
80.2	80.1
79.8	79.4

Calculate the observed significance level for the hypothesis that there is no difference in the two methods of instruction.
7. Two members of a coffee club become involved in a dispute about the shape of coffee cups. One person claims that there is more heat loss from cups which flare outward at the top than from cups which are regular cylinders. They design a small experiment to compare the two shapes. Hot coffee is poured in five cups, two flared and three not.

They measure the heat loss (°F) over a 10-minute period. The results are:

Flare	No Flare
22	18
24	23
	19

Estimate the mean difference in heat loss for the two populations of cups and estimate the variance of this estimate. Calculate the observed significance level for the hypothesis of no difference.

GEORGE W. SNEDECOR. 1882–1974. Reproduced by permission from the Department of Statistics, Iowa State University.

27

THE ANALYSIS OF VARIANCE

27.1 The **analysis of variance** was published in 1923 by Fisher [1] just 5 years after he used the word **variance** to describe the square of the standard deviation [2]. Although Fisher is regarded as the originator of the analysis of variance, some of the ideas certainly predate him. Stigler [3] calls attention to the fact that Edgeworth carried out calculations much like the analysis of variance as early as 1885. Since 1923 the analysis of variance has become an extremely useful statistical analysis (perhaps more frequently performed than any other). Countless articles and books have been written concerning some facet of the analysis of variance, and it has been theoretically described at virtually every level of mathematics. In this lecture we wish to give some insight into the nature of the analysis of variance at an introductory level, using relatively little mathematics. However, it is felt that some arithmetic is unavoidable.

27.2 In fact, the analysis of variance is not a study of variance exactly but of **variation**. For a concise, intuitive statement we commend the description of Snedecor [4]. "It is a technique for segregating from comparable groups of data the variation traceable to specified sources."* This simple statement gives a good idea of what we are about and is a beacon to prevent our getting lost in computational details.

Once we decide to try to segregate variation in our data by specified **sources** of variation, we must decide how we are going to **measure** variation. In fact, the only way of measuring variation that has proven useful for this purpose is the **sum of squares** of deviations from the sample

* Reprinted by permission from *The Iowa State University Press* from Snedecor, *Calculation and Interpretation of Variance and Covariance*, Snedecor, 1934.

mean. The analysis of variance, then, is concerned with **partitioning** the **total variation** in the data (as measured by the sum of squares of deviations from the sample mean) into **bits of variation** (also measured by sums of squares) traceable to **specified sources**.

27.3 For the first example of an analysis of variance, let us return to the example in Lecture 23. The first exam scores (X) were used to describe the final exam scores (Y). The predicted final exam scores ($\hat{Y}$) were the final exam scores predicted by the least squares line. The data are summarized in Table 27.1.

In the spirit of Snedecor's description we wish to partition and segregate the variation traceable to specified sources. We have already calculated the sum of squares of deviations from the sample mean to be

$$\Sigma(Y - \bar{Y})^2 = 341.875$$

What are the possible sources for this variation? Obviously, one reason that there is variation in the Y values is the apparent **linear trend.** Let us calculate the sum of squares of deviations from the sample mean of the $\hat{Y}$ values.

$$\Sigma \hat{Y}^2 = 48{,}437.1529$$

and

$$\Sigma \hat{Y} = \Sigma Y = 621$$

so

$$\Sigma(\hat{Y} - \bar{Y})^2 = 48{,}437.1529 - 48{,}205.125$$
$$= 232.0279$$

Table 27.1 Regression Example

	X	Y	$\hat{Y}$
	63	68	69.9749
	65	75	71.0581
	72	70	74.8493
	73	76	75.3909
	80	81	79.1821
	85	78	81.8901
	86	89	82.4317
	93	84	86.2229
Total	617	621	621.0000

The other source or reason for variation is that not all of the Y values fall on the regression line. We measure this variation by Σd^2, where d represents the **deviation** of the observed Y value from the predicted value $\hat{Y}$. That is, $d = Y - \hat{Y}$ and $\Sigma d^2 = \Sigma(Y - \hat{Y})^2$. In Lecture 23 we calculated

$$\Sigma d^2 = 109.8471$$

No other sources of variation occur to us and we note that

$$341.875 = 232.0279 + 109.8471$$

or

total SS = regression SS + residual SS,

where SS = sum of squares of deviations from the sample mean. It is conventional to record these results (along with other quantities to be explained shortly) in a table similar to Table 27.2. Notice that AOV is used as an abbreviation for analysis of variance.

The columns headed **Source** and **SS** should be clear. However, the columns headed **df** and **MS** require further explanation. **Degrees of freedom**, abbreviated df, are difficult to explain at the beginning, and the name seems strange. The name and meaning are related to the use of the expression in engineering and physics. Beginning students are perhaps well advised simply to accept the idea and terminology at first. Use and familiarity come to give meaning to degrees of freedom. The abbreviation MS means **mean square**, and the entries in this column are obtained by dividing the sum of squares by their appropriate degrees of freedom.

The immediate use that can be made of the mean square column (given certain assumptions) is to perform an ***F* test** of the hypothesis that the regression coefficient is zero.

$$F = \text{regression MS/residual MS}$$

$$= 232.0279/18.3078$$

$$= 12.6737$$

Table 27.2 Regression AOV

Source	df	SS	MS
Total	7	341.8750	
Regression	1	232.0279	232.0279
Residual	6	109.8471	18.3078

From Table A.7, we see that the probability of an F this large with 1 degree of freedom for the numerator and 6 degrees of freedom for the denominator is about .01. That is,

$$\text{SL} = P(F \geq F \text{ observed}) = .01$$

Therefore we feel strongly that $\beta_1 \neq 0$.

27.4 For those with more algebraic inclinations the partition of the total sum of squares in Table 27.2 is simply explained by the following algebraic identity.

$$\begin{aligned}\Sigma(Y_i - \bar{Y})^2 &= \Sigma(Y_i - \hat{Y}_i + \hat{Y}_i - \bar{Y})^2 \\ &= \Sigma(\hat{Y}_i - \bar{Y})^2 + \Sigma(Y_i - \hat{Y}_i)^2 \\ &= \textbf{regression SS} + \textbf{residual SS}\end{aligned}$$

27.5 For the next example of an analysis of variance consider data that are grouped into several classes [e.g., (1) family incomes in several different cities, (2) ages of people using different libraries, and (3) IQs of people in different university classes]. Unlike the example of the preceding sections, there is no regression line and the recognizable sources of variation are simply **between classes** and **within classes**. In order to keep the arithmetic simple, let us consider another small example. Suppose that the data in Table 27.3 represent the weight gains of 12 animals in a nutrition experiment, with 4 animals receiving one of three possible rations.

Table 27.3 Nutrition Experiment

	Ration								
	A			B			C		
	Y	$\hat{Y}$	$(Y-\hat{Y})^2$	Y	$\hat{Y}$	$(Y-\hat{Y})^2$	Y	$\hat{Y}$	$(Y-\hat{Y})^2$
	24	27	9	24	29	25	25	27.75	7.5625
	30	27	9	26	29	9	30	27.75	5.0625
	26	27	1	32	29	9	27	27.75	0.5625
	28	27	1	34	29	25	29	27.75	1.5625
Total	108		20	116		68	111		14.7500
Mean	27			29			27.75		

In Table 27.3 the weight gains are represented by Ys, and the other columns have been constructed in a manner analogous to that for the regression example. For ration A the sample mean 27 is taken as the predicted value $\hat{Y}$, 29 as $\hat{Y}$ for ration B, and 27.75 for ration C. Then the column $(Y - \hat{Y})^2$ follows in a natural manner. For this example,

$$\begin{aligned}\Sigma(Y - \bar{Y})^2 &= \Sigma Y^2 - (\Sigma Y)^2/12 \\ &= 9463 - 9352.0833 \\ &= 110.9167\end{aligned}$$

Using the same notation as for the regression example,

$$\begin{aligned}\Sigma(\hat{Y} - \bar{Y})^2 &= \Sigma \hat{Y}^2 - 9352.0833 \\ &= 4(27)^2 + 4(29)^2 + 4(27.75)^2 \\ &\quad - 9352.0833 \\ &= 8.1667\end{aligned}$$

$$\begin{aligned}\text{Then the residual SS} &= \Sigma(Y - \hat{Y})^2 \\ &= 20 + 68 + 14.75 \\ &= 102.75\end{aligned}$$

Assembling these quantities in an analysis of variance table, we have Table 27.4.

The degrees of freedom are arrived at in the following manner. The **total degrees of freedom** are one less than the number of observations, the **between degrees of freedom** are one less than the number of groups or classes, and the **within degrees of freedom** are obtained by subtracting. Alternatively, within degrees of freedom are obtained by adding the degrees of freedom within groups (3 + 3 + 3).

Table 27.4 One-Way AOV

Source	df	SS	MS
Total	11	110.9167	
Between	2	8.1667	4.0834
Within	9	102.7500	11.4167

The F test for testing equality of ration population means is simply

$$F = \frac{\textit{between MS}}{\textit{within MS}}$$

$$= \frac{4.0834}{11.4167} < 1$$

From Table A.7 we see that the probability of an F this large is extremely great, so the hypothesis of equal means is quite acceptable.

27.6 The analysis of variance of the **one-way classification** just given can be described simply by an algebraic identity. With n observations in each of t groups, let Y_{ij} denote the jth observation in the ith group, M_i the mean of the ith group, and M the overall mean. Then

$$\sum_{ij} (Y_{ij} - M)^2 = \sum_{i} n_i(M_i - M)^2 + \sum_{ij} (Y_{ij} - M_i)^2$$

27.7 We have introduced the analysis of variance as a technique for **partitioning** the **total variation** into **recognizable sources** of **variation**. Our illustrations cover only simple linear regression and the one-way classification, so that there are only three sources of variation. Obviously, the full power is realized only when there are more sources of variation. Nevertheless, the ideas in more complicated situations are similar to the ideas expressed in this lecture.

27.8 Fisher gave the analysis of variance in his *Statistical Methods for Research Workers*. However, the table he gave was for z, not F. This required taking one-half the natural logarithm of the ratio of mean squares. This awkward calculation was eliminated by the tabulation of $F = e^{2z}$. Although the F table was first tabulated by Mahalanobis in 1932, the F became widely known as Snedecor's F. The F tables appear in Snedecor's book [4] that was published in 1934. Snedecor says simply that the table was computed from Fisher's z table. Despite the obvious advantage of F over z, Fisher continued to give the z tables in subsequent editions of his *Statistical Methods*.

SUMMARY. The **analysis of variance,** first given by Fisher in 1923, is one of the most frequently performed statistical **analyses**. Basically, it is an analysis not of variance but of **variation** as measured by the sum of squares of deviations from the sample mean. This sum of squares, called the **total sum of squares**, is partitioned into pieces traceable to recognizable **sources** of variation.

With simple linear regression, the total sum of squares is partitioned into two terms: the sum of squares due to **regression** and the **residual** sum of squares due to deviations from the regression line. The residual SS can be written as residual SS = (total SS)$(1 - r^2)$.

With a one-way classification of data, the total SS is partitioned into the **between groups** SS and the **within groups** SS.

Each sum of squares has certain degrees of freedom and a **mean square** is the sum of squares divided by its degrees of freedom. To test equality of group means one uses an ***F* statistic**, which is the ratio of between groups mean square to the within groups mean square.

REFERENCES

1. Fisher, R. A., and W. A. Mackenzie. (1923). "Studies in Crop Variation, II. The Manurial Response of Different Potato Varieties," *Journal of Agricultural Science*, *13*, 311–320.
2. Fisher, R. A. (1918). "The Correlation Between Relatives on the Supposition of the Mendelian Inheritance," *Transactions of the Royal Society of Edinburgh*, *52*, 399–434.
3. Stigler, Stephen. (1978). "Francis Ysidro Edgeworth, Statistician," *Journal of the Royal Statistical Society*, *A-141*, 287–322.
4. Snedecor, George W. (1934). *Calculation and Interpretation of Analysis of Variance and Covariance*, Ames, Ia.: Collegiate Press.

EXERCISES

1. Verify the analysis of variance in Table 27.2
2. Verify the analysis of variance in Table 27.4.
3. Prove the identity in Section 27.6.
4. Construct an analysis of variance table such as Table 27.2 for the data of Exercise 1 in Lecture 23. Using the F statistic, calculate the significance level for $H_0: \beta_1 = 0$ versus $H_1: \beta_1 \neq 0$.
5. Repeat Exercise 4 for the data of Exercise 2 in Lecture 23.
6. A researcher in child psychology designed an experiment that required children to push a button when they recognized pictures

flashed on a screen. He measured the reaction times in seconds for children in four age groups.

4-year	5-year	6-year	7-year
40	37	34	32
42	36	31	28
38	35	32	27
46	42	38	33
41	48	37	29

Construct an analysis of variance for these data similar to Table 27.4. Using the F statistic, calculate the significance level to test the equality of reaction times for the four age groups.

7. Faculty members responding to a salary survey gave their rank, salaries, and years in rank. The years in rank, X, and annual salaries, Y, in thousands of dollars, follow.

X	Y	X	Y
5	25.5	20	42.1
10	27.5	12	29.4
8	28.3	16	33.4
15	32.5	8	27.8
18	36.3	2	23.1

Calculate the analysis of variance as in Table 27.2 and the F statistic. What is the significance level for the hypothesis that the regression coefficient is zero?

8. The cooking times for green beans reported by housewives over the country at different altitudes were as shown in the following table. Calculate the analysis of variance table and F statistic. Calculate the significance level for the hypothesis that altitude does not affect the cooking time. Altitude in thousands of feet $= X$ and cooking time in minutes $= Y$.

X	Y	X	Y
.6	25	6.0	58
.8	30	3.0	42
1.2	35	1.1	33
2.2	41	.5	20
3.6	46	.4	15
5.0	54	2.0	40

9. A sociologist, studying attitudes on religious issues, administers a questionnaire to university students. Each student responds to 50 statements on a 1 to 5 scale, the higher response being a more conservative response. Then the average response for 50 items is recorded for each student. The students fall into three groups: no religious training, some religious training, considerable religious training.

None	Some	Considerable
4.2	2.4	4.6
4.0	1.8	3.9
3.1	2.2	3.5
2.7	1.9	3.7
2.3	2.4	2.8
3.3	2.7	3.4
4.1		3.6
		3.8

Perform the analysis of variance and calculate the F statistic and the significance level for the hypothesis of no differences among the group means.

10. At the twenty-fifth reunion of a high school graduating class, the participants completed an information sheet that gave their annual salary and highest university degree earned. Two responses that were very high were discarded. The remaining responses produced the following statistics.

	n	$\bar{X}$	s
B.S.	65	17,250	2,125
M.A.	12	21,125	3,275
Ph.D.	8	26,340	2,960

Perform the analysis of variance, calculate the F statistic and calculate the significance level for the hypothesis that the mean salary is the same for all three groups.

SOME RANDOMIZED EXPERIMENTS

EXPERIMENTAL DESIGN. A forestry experiment begun in 1929 and photographed about 1945. Courtesy of The Forestry Commission, England.

28

THE CONCEPT OF EXPERIMENTAL DESIGN

28.1 The world of experimental science is indebted to Fisher for his pioneer work at Rothamsted. Most certainly experimental science did not begin with Fisher but **experimental design**, as we now use the term, did. Fisher's article [3] paved the way for his work of monumental importance, *The Design of Experiments* [4]. The ideas that flowed from that work have been widely disseminated by books such as Snedecor's *Statistical Methods* [6] and *Experimental Design* by Cochran and Cox [1]. The well-known books by Kempthorne [5] and Cox [2] are at a more advanced level.

28.2 In a way the expression **experimental design** is misleading. We are not concerned with the design of laboratory equipment or the breeding of colonies of laboratory animals such as mice or birds, nor are we directly concerned with the design of instruments to measure experimental responses. Instead, we are concerned with the **logic** of collecting and analyzing data. By experimental design we mean the **plan** for generating the data so that the analysis can use methods based on well-developed theory, tested and tried by one's peers.

28.3 It must also be emphasized that this lecture is concerned with a certain type of experiment—a **comparative** experiment, not an **absolute** experiment. An absolute experiment is conducted to determine the absolute magnitude of a physical constant or the absolute existence or non-existence of some phenomenon. A comparative experiment, on the other hand, is conducted to compare the relative values of several constants and is concerned with the differences, or perhaps the ratios, among them.

For example, one successful transmission of the human voice by wireless was enough to establish absolutely that radio programs were possible. On the other hand, many comparative experiments were conducted in order to arrive at today's design for transmitters and receivers.

28.4 Next we present a certain way of describing a comparative experiment. Although it will not fit all cases, it describes most of them and has proven extremely useful. It might even be called the **canonical form** of a comparative experiment. We consider the case where **stimuli** are applied to some sort of **experimental material**, producing a measurable response. The stimuli are called **treatments**, a generic term only, and the experimental material is organized into units called **experimental units**. We will say more about the definition of an experimental unit. Measurement of the responses gives **observations** that are then **analyzed**. Finally, the statistical analysis must be **interpreted** by the experimenter. These ideas are portrayed in Figure 28.1. Although an experiment is designed with an analysis in mind, the design is determined by the way in which the **experimental material** is **organized** and the way in which the **treatments** are **applied** to the **experimental material**.

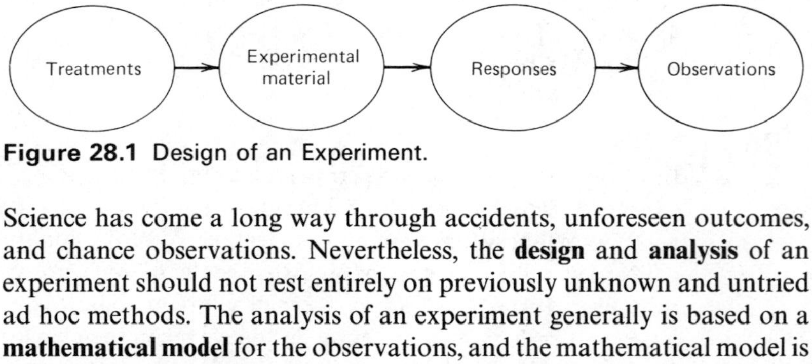

Figure 28.1 Design of an Experiment.

28.5 Science has come a long way through accidents, unforeseen outcomes, and chance observations. Nevertheless, the **design** and **analysis** of an experiment should not rest entirely on previously unknown and untried ad hoc methods. The analysis of an experiment generally is based on a **mathematical model** for the observations, and the mathematical model is related to and justified by the design. In the lecture on randomization we emphasized that the physical act of randomization justifies considering the conceptual population of different assignments of the treatments. It might even be said that randomization is the step that introduces probability models for the observations. This fact was perceived by Fisher and must be regarded as a fundamental contribution.

Frequently probability models using the normal distribution are introduced into the analysis. The link between randomization and normally distributed observations is not well defined, but it seems to be justified on the basis of considerable experience.

28.6 In Section 28.3 we referred briefly to the need for more observations in comparative experiments than in absolute experiments. It is more than a need for several observations. We must have several independent randomizations of treatments to experimental units. Such independent randomizations are called **replications**. To illustrate the idea of replication, consider designing an experiment to compare the effects of three different plant growth stimulants on green bean plants grown in a greenhouse. Little credence would be attached to an experiment in which treatments A, B, and C were assigned at random to one plant each. The relative merits of A, B, and C would be dictated by the relative merits of the three plants to which they were assigned. On the other hand, if the three treatments are randomly assigned to many plants, we feel that the plant differences will "average out," and the relative effects of A, B, and C will have a chance to manifest themselves.

Many observations do **not** necessarily mean that there have been many replications. Suppose that the plants in the greenhouse are arranged on three tables and the treatments are randomly assigned to the three tables, with all plants on a given table receiving one treatment. Although this would result in many observations, there is only one replication, and it has the same status as the experiment involving three plants only. As far as comparison of A, B, and C is concerned, it would yield only three numbers, the total responses for the three tables involved.

28.7 The comparison of treatments ultimately involves comparing two treatments, so let us consider the problem of only two treatments. We will calculate from our data an estimate of the difference between the two treatments effects, say $\hat{\tau}$. This estimate will be in error. That is,

$$\hat{\tau} = \tau + e$$

or **estimate** = **true effect** + **experimental error**. What does the experimental error include? Stated another way, what factors cause our estimate to be anything other than the true effect? The following list has been useful in trying to answer this question.

1. Differences in experimental units.
2. Errors in applying treatments.
3. Errors in measuring the response.
4. Unknown factors.
5. Factors known but ignored.

All of these factors contribute to the experimental error.

In judging the significance of our estimate $\hat{\tau}$, we need to estimate the variance of our experimental error (called briefly **experimental error variance**). It seems almost obvious that any estimate of experimental error variance should include **all** of the factors that produced the experimental error in the first place. For example, in the greenhouse experiment with the treatments assigned at random to the tables, the experimental error variance cannot be estimated by differences among plants on the same table because these differences do not include table differences.

28.8 In designing experiments it is frequently the case that we look ahead to the analysis of variance likely to result from data. This is so because the analysis of variance is so often used for data from designed experiments. Frequently some writers may refer to the analysis of variance or to the model on which it is based as the design. However, we reserve the word **design** to describe the plan for assigning the treatments to the experimental material.

SUMMARY. The ideas of **experimental design** draw heavily on the work of Fisher. In **comparative** experiments the object is to compare two or more stimuli called **treatments**. The treatments are assigned to **units** of **experimental** material, which results in responses and **observations**. The observations are then analyzed and the analysis interpreted. The design of the experiment is concerned primarily with the organization of the **experimental material** and the **assignment** of the **treatments**.

Analysis of the data results in an estimate $\hat{\tau}$ of the treatment effect τ, but the estimate is in error. Much effort is concerned with estimating the variance of this experimental error. **Replication** (i.e., independent repetitions of treatments) has the effect of decreasing the experimental error variance. At the same time, it makes possible estimation of the experimental error variance.

REFERENCES

1. Cochran, W. G., and Gertrude Cox. (1957). *Experimental Designs*, 2nd Edition, New York: Wiley.
2. Cox, D. R. (1958). *Planning of Experiments*, New York: Wiley.
3. Fisher, R. A. (1926). The Arrangements of Field Experiments," *Journal of the Ministry of Agriculture* (England), *33*, 503–513.

4. Fisher, R. A. (1935). *The Design of Experiments* (7th Edition, 1960), Edinburgh: Oliver & Boyd.
5. Kempthorne, O. (1952). *The Design and Analysis of Experiments*, New York: Wiley.
6. Snedecor, George W. (1967). *Statistical Methods*, 6th Edition, with W. G. Cochran, Ames: Iowa State Press.

THE SALK VACCINE. A nationwide immunization program based in part on a two-group experiment. Wide World.

29

TWO-GROUP AND PAIRED EXPERIMENTS

29.1 In 1954 a public experiment of massive scale was conducted in the United States to evaluate the effectiveness of the Salk vaccine. Meier [2] gives an excellent account of the role that statistics played in the Salk vaccine field trials. Involving over 1 million children at a total direct cost of $5 million, the trials needed planning according to sound principles of experimental design. Part of the study was planned as a **two-group design**. About 200,000 children were inoculated with Salk vaccine; another 200,000 children were inoculated with a control (a "sugar water" that resembled the Salk vaccine in appearance). The polio rate for these two groups was markedly **different**. Eighty-two of the vaccinated contracted polio; 162 of those who received the control contracted polio. This demonstration of the effectiveness of the vaccine helped lead to its general usage.

A much smaller two-group experiment is described by Pfeiffer [3, p. 86]. From the time of Aristotle, scientists had accepted the idea that certain forms of life were generated when other matter decayed. This idea was challenged in 1668 by Francesco Redi. Redi put meat in two jars, covering one and leaving the other uncovered. Meat in both jars decomposed, but only the open jar generated flies. This experiment disproved the idea of spontaneous generation of life from decaying matter.

We now wish to describe in detail **two-group** and **paired** experiments. To simplify calculations, we will consider small-scale experiments.

How would you design an experiment to compare the effectiveness of two headache remedies? Both of the remedies, *A* and *B*, are advertised by their manufacturers as being superior to the other. We are interested in only one aspect of effectiveness: time required for relief from pain.

Suppose that all of the 20 students in your statistics class have volunteered to participate in whatever experiment you design.

In this case one of the major difficulties is in measuring time to relief because what constitutes relief from pain for one person presumably will not for another. Also, people will feel the need to take headache remedies at different thresholds of pain. After studying the matter of pain and reviewing the literature on similar experiments, you decide to proceed as follows. When one of your volunteers experiences a headache, he or she will take the prescribed remedy when the pain becomes severe and record the time required before becoming comfortable.

29.2 Having devised a measure of time to relief, you must now decide which randomized design to use: a **two-group** design or a **paired** design. The **two-group** design would consist of selecting some of the 20 volunteers at random to receive remedy A; the others would receive remedy B. Generally, we would select 10 for A and 10 for B. The **paired** design calls for matching or pairing the subjects into 10 pairs so that the individuals in each pair are as much alike as possible and then, for each pair, randomly assigning A to one individual and B to the other individual. If there is a sound basis for pairing, the idea is that within each pair we will get a very precise comparison of A with B, and we can then combine all 10 of these precise comparisons. If, on the other hand, there seems to be no basis for pairing, the two-group design would likely be preferred because of its simplicity. In the problem under discussion you would need to have considerable information about reaction to pain of your volunteers before you could expect to pair them intelligently. We will continue this lecture by following through examples for both designs being discussed.

29.3 If you decide to use the **two-group** design, you do so knowing before observations are ever collected how you are going to analyze the data. The data can be analyzed using calculations such as those illustrated in the randomization lecture. More commonly, the analysis will be based on the **model** that observations A and B represent two random samples from two normal populations. The analysis will be directed toward forming opinions about the means and variances of these populations. In what follows we will illustrate the analysis of the data in Table 29.1, assuming that the two population variances are equal and that we wish to compare the means.

Table 29.1 Data from Two-Group Experiment Relief Times (Minutes) from A and B

	A	A^2	B	B^2
	25	625	32	1024
	16	256	44	1936
	20	400	46	2116
	30	900	28	784
	46	2116	36	1296
	22	484	48	2304
	83	6889	53	2809
	26	676	76	5776
	34	1156	58	3364
	26	676	42	1764
Total	328	14,178	463	23,173
Mean	32.8		46.3	

The relief times are given in columns A and B. The squares of these numbers are given in columns A^2 and B^2. Before performing any other calculations, note the average relief times for A and B. Remedy A produced relief 13.5 minutes faster on the average, so one might reasonably ask why further calculations are necessary. Isn't it obvious that the experiment proves A is better than B? But wait! Among those individuals receiving remedy A and among those individuals receiving B there are differences larger than 13.5. The proper approach is to try to judge whether the difference of 13.5 in the means is significant when we consider the differences among experimental units treated alike. Put another way, we are interested in the probability of a difference as large as 13.5 occurring by chance even if the population means for A and B are equal. To find this probability, we will estimate the standard deviation of the difference of means and calculate a t value.

$$\bar{A} = 32.8 \qquad \bar{B} = 46.3$$

$$(\Sigma A)^2/10 = 10{,}758.4 \qquad (\Sigma B)^2/10 = 21{,}436.9$$

$$\Sigma(A - \bar{A})^2 = \Sigma A^2 - (\Sigma A)^2/10 \qquad \Sigma(B - \bar{B})^2 = \Sigma B^2 - (\Sigma B)^2/10$$

$$= 14{,}178 - 10{,}758.4 \qquad = 23{,}173 - 21{,}436.9$$

$$= 3419.6 \qquad = 1736.1$$

Assuming that the two populations have the same variance, it seems reasonable to use the variation from both samples to estimate the common variance. This estimate is given by **pooling** (or adding) the sum of squares of deviations about the sample mean and dividing by the pooled (or added) degrees of freedom. That is,

$$s^2 = \frac{\text{pooled SS}}{\text{pooled df}}$$

$$= \frac{3419.6 + 1736.1}{9 + 9}$$

$$= 286.4278$$

Now the variances of the sample means are estimated by s^2/n_A and s^2/n_B. Because the variance of a difference of means (as well as the sum) is the sum of variances, the variance of the mean difference $\bar{A} - \bar{B}$ is estimated by

$$s^2{}_{\bar{A}-\bar{B}} = s^2(1/n_A + 1/n_B)$$

$$= (286.4278)(1/10 + 1/10)$$

$$= 57.2856$$

Then the standard deviation for $\bar{A} - \bar{B}$ is simply the square root of the estimated variance. This gives

$$s_{\bar{A}-\bar{B}} = 57.2856$$

$$= 7.57$$

Finally, the t to test the hypothesis of no treatment difference $H_0: \mu_A - \mu_B = 0$ versus the alternative of a treatment difference $H_A: \mu_A \neq \mu_B$ is simply

$$t = \frac{\bar{A} - \bar{B} - 0}{s_{\bar{A}-\bar{B}}}$$

$$= -13.5/7.57$$

$$= -1.78$$

If the hypothesized difference had been something other than zero (say 3), then 3 and not zero would have been subtracted from $\bar{A} - \bar{B}$. The final step in this test of significance is to calculate the probability of a t larger than this (in absolute magnitude). This probability or **significance level**, SL, is given by

$$SL = P(|t| \geq 1.78)$$

which, from Table A.5, is about .10. Thus the difference of 13.5 is a little unlikely if the population means were equal, but it is not compellingly in favor of unequal means, as we might have surmised.

We can graph values of SL versus various hypothesized values of $\mu_A - \mu_B$ to form a more complete picture of tenable values for the population mean difference. From such a graph we would find that all values of $\mu_A - \mu_B$ between

$$\bar{A} - \bar{B} - 2.101 s_{\bar{A}-\bar{B}} \quad \text{and} \quad \bar{A} - \bar{B} + 2.101 s_{\bar{A}-\bar{B}}$$

or

$$-29.4 \quad \text{and} \quad 2.4$$

will produce a SL greater than .05. We might then say that all such values are in **consonance** with (in agreement with) the data at the .95 level. Such an interval is called a **consonance interval** by Kempthorne and Folks [1]. Here 2.101 is simply the tabulated t value with 18 degrees of freedom such that the probability of a larger value in absolute magnitude is .05.

29.4 Next we will illustrate the analysis using a paired **design.** We will not give as much discussion at each step. The idea of a significance test is the same: we calculate a t value and then determine the probability of a larger value in absolute magnitude. In Table 29.2 we give results that might have been obtained under columns A and B.

Remember that one of the reasons for doing a paired experiment is to get a **precise comparison** of treatments **within** each pair. The first step is then to take the difference between relief time for each pair. In the example we have calculated $D = B - A$. In fact, we could have calculated $A - B$ for each pair. It does not matter which we do, as long as we do the same for all pairs. The **model** we now use as basis for further analysis is that the 10 differences (Ds) comprise a random sample from a normal population with mean $\mu_D = \mu_B - \mu_A$ and variance $\sigma_D{}^2$. We want to perform a test of significance for $H_0\colon \mu_D = 0$ versus $H_A\colon \mu_D \neq 0$.

Table 29.2 Data from Paired Experiment

Pair	A	B	$D = B - A$	D^2
1	36	37	1	1
2	44	48	4	16
3	61	72	11	121
4	26	38	12	144
5	32	38	6	36
6	33	30	−3	9
7	45	58	13	169
8	42	48	6	36
9	56	63	7	49
10	53	58	5	25
Total			62	606

The variance $\sigma_D{}^2$ is estimated by $s_D{}^2 = \Sigma(D - \bar{D})^2/(n - 1)$, where n is the number of pairs. Now

$$\Sigma(D - \bar{D})^2 = \Sigma D^2 - (\Sigma D)^2/10$$
$$= 606 - 384.4$$
$$= 221.6$$

so

$$s_D{}^2 = 221.6/9$$
$$= 24.6222$$

and

$$s_D = 4.96$$

The variance of $\bar{D}$ is estimated by

$$s_{\bar{D}}{}^2 = s_D{}^2/n$$
$$= 2.4622$$

with

$$s_{\bar{D}} = \sqrt{2.4622}$$
$$= 1.5691$$

The t value is then given by

$$t = \frac{\bar{D} - 0}{s_{\bar{D}}}$$

$$= \frac{6.2}{1.5691}$$

$$= 3.951$$

(For $H_0: \mu_D = c$ we would subtract c.) From Table A.5 we determine (using 9 df) that

$$\text{SL} = P(|t| \geq 3.951) \doteqdot .01$$

A (.95) consonance interval is given by

$$D \pm 2.262 s_{\bar{D}}$$

or

$$6.2 \pm 3.55$$

or

$$2.65 \text{ and } 9.75$$

Thus values of μ_D between 2.65 and 9.75 would yield significance levels greater than .05.

29.5 A striking illustration of a **two-group** experiment is given by data from part of an experiment performed by Wakeman and Kaplan [5]. Burn patients were assigned to one of two groups: **medication only** and **medication under hypnosis**. The percent of allowable medication was recorded for each patient. For four patients receiving hypnosis the average percent was 40. For five patients receiving medication only the average percent was 80. Thus the patients receiving hypnosis required a much lower level of medication.

29.6 The results of an interesting **paired** experiment are reported by Sackeim, Gur, and Saucy [4]. Pictures of **human faces** displaying different emotions were split down the middle and **left-side** and **right-side composites** were constructed. The right-side composite was formed by taking the right side of the original face and forming a mirror image to form the left side. Similarly, the left-side composite was a symmetric face with the right side being the mirror image of the left side of the original face. Subjects were then shown slides of left- and right-side composites and asked to rate on a 1 to 7 scale the intensity of expression. The **conclusions** of the experiment are indicated by the "Emotions Are Expressed More Intensely on the Left Side of the Face."

SUMMARY. Two of the simplest **experimental designs** are the **paired** and **two-group** designs. The paired design assigns two treatments to experimental units at random **within** each pair of experimental units. If there are n pairs, there are n independent **random assignments**. The two-group design assigns one treatment to some of the experimental units chosen at random and the other treatment to the other experimental units. Thus, with the two-group experiment, there is only **one random assignment**.

The paired experiment results in n pairs of observations, and the analysis is performed on the n **differences**. The **model** for the analysis is that the n differences are a random sample from a single normal distribution.

The two-group experiment results in two samples of observations, and the model for the analysis usually is that we have independent random samples from two normal populations, with equal variances but different means, possibly.

Several examples are given of both types of experiments.

REFERENCES

1. Kempthorne, O., and L. Folks. (1971). *Probability, Statistics, and Data Analysis*, Ames: Iowa State Press.
2. Meier, Paul. (1972). "The Biggest Public Health Experiment Ever: The 1954 Field Trial of the Salk Poliomyelitis Vaccine," in *Statistics, A Guide to the Unknown*, edited by Judith M. Tanur, et al., San Francisco: Holden-Day.
3. Pfeiffer, John. (1964). *The Cell*, New York: Time.
4. Sackeim, H. A., R. C. Gur, and M. C. Saucy. (1978). "Emotions Are Expressed More Intensely on the Left Side of the Face," *Science, 202*, 434–436.
5. Wakeman, R. John, and Jerold Z. Kaplan. (1978). "An Experimental Study of Hypnosis in Painful Burns," *The American Journal of Clinical Hypnosis, 21*, 3–12.

EXERCISES

1. Verify the calculations in Section 29.3.
2. Verify the calculations in Section 29.4.
3. There is often confusion among beginning students of statistics about whether two groups of data should be regarded as paired or unpaired. From the viewpoint put forward in this lecture, how can we tell? What is the essential difference in randomization between a paired experiment and a two-group experiment?
4. Wakeman and Kaplan do not give the basic data in their paper, only the averages. Suppose that the percents of allowable medication requested by nine patients are as follows.

Medication Only	Hypnosis/ Medication
85	26
92	46
76	37
81	51
66	

Calculate the t statistic to compare means and determine the observed significance level for the null hypothesis that $\mu_1 = \mu_2$ versus the alternative hypothesis that $\mu_1 \neq \mu_2$.

5. Sackheim, Gur, and Saucy do not give their raw data; they report only the conclusions. Suppose that we submitted 10 each of left- and right-side composites to a panel of eight judges. Each judge rated each composite on a 1 to 7 scale. Then 10 pairs of average ratings are as follows.

Right-Side Composite	Left-Side Composite
3.375	3.625
4.750	4.875
5.125	5.250
6.250	6.125
3.750	4.125
4.875	5.250
5.125	5.875
6.250	6.625
5.875	6.250
5.250	6.375

Calculate the t statistic for comparing means. Determine the observed significance level for testing $H_0: \mu_R = \mu_L$ versus $H_1: \mu_R \neq \mu_L$.

6. Cub scouts build and race small model cars in a derby. At a local elimination meet, the race track had two lanes. The boys felt that the left lane was faster than the right lane, so some trial runs were made with each car running in both the left and right lanes. The running times were clocked with a stopwatch, and the times are as follows.

Car	Left	Right
1	5.1	5.3
2	5.6	5.4
3	4.9	5.2
4	4.8	5.2
5	5.3	5.4
6	5.6	5.5
7	5.6	5.6
8	5.3	5.2
9	5.1	5.3

Calculate the t statistic for comparing means. Determine the observed significance level for testing $H_0: \mu_L = \mu_R$ versus $H_1: \mu_L < \mu_R$.

7. Each member of an 18-member taste panel is presented with two pieces of sirloin steak in random order. One of the pieces has been aged for 6 months before cooking; the other is from recently slaughtered beef. Each member of the panel rates the two pieces of beef on a five-point scale. The ratings follow.

Aged	Fresh	Aged	Fresh	Aged	Fresh
4	3	5	1	3	3
4	2	5	2	4	3
5	2	5	3	4	2
4	4	5	4	2	2
5	3	3	2	3	1
3	1	4	2	4	3

Test the hypothesis that there is no difference between the mean ratings versus the hypothesis that there is. Use $\alpha = .05$.

8. A restaurant is considering two different brands of casual china. One

concern is how well the coffee cups hold the heat. A two-group experiment is designed to compare the heat losses from the two brands of china cups. The statistics from the experiment are:

	n	$\bar{X}$	SS
Brand A	20	18.2	82.62
Brand B	20	22.3	78.96

Test $H_0: \mu_A = \mu_B$ versus $H_A: \mu_A \neq \mu_B$. Use $\alpha = .01$.

9. A researcher in business administration is concerned with the effects on office efficiency of the cocktail for lunch. A two-group experiment is used. Some typists selected at random from a group of volunteers are given 1 ounce of alcohol during lunch. Others are given 2 ounces. During the first hour after lunch, the typing speed and error rate is recorded for each participating typist. Calculate the observed significance level for the hypothesis of no difference in mean typing speeds and in error rates.

		Typing Speed		Error Rate	
	n	$\bar{X}$	SS	$\bar{X}$	SS
One ounce	12	55	24,122	5.2	76.41
Two ounces	14	62	26,304	7.4	82.33

10. In a large department each instructor is assigned one lower-division course and one upper-division course. At the end of the term, student evaluations are administered for all courses. This results in a pair of overall ratings for each instructor. Test the null hypothesis that there is no difference between student evaluation ratings for upper- and lower-division courses. Use $\alpha = .05$.

Upper	Lower
3.21	2.93
3.41	3.12
2.98	2.86
3.35	2.68
3.67	3.25
3.24	3.01
3.13	2.94

COMPLETELY RANDOMIZED DESIGN. Wisconsin fields. USDA.

30

COMPLETELY RANDOMIZED AND RANDOMIZED BLOCK DESIGNS

30.1 In this lecture we illustrate two very simple (but much used) designs: the **completely randomized** (or completely random) and the **randomized block** designs. As in the last lecture, we will illustrate the analysis using small sets of artificial data in order to keep the arithmetic simple. After a discussion of the planning of the experiment, we will illustrate the arithmetic required to perform an **analysis of variance** and a subsequent ***F* test of significance**.

30.2 Appreciation for the design of experiments may be more easily acquired if one is actually involved in a research situation or if one is well acquainted with the research of others. Lectures 28, 29, and 30 may seem rather abstract to a student who has never seen or who has never been involved with a research project. Therefore it might be appropriate to compare the effect on student final scores, in a course in statistics, of three different methods of teaching.

1. **Method A.** Regular statistics lectures supplemented with handout materials describing real examples from current research.
2. **Method B.** Same as method *A* with the additional requirement that the students design, conduct, and analyze data from experiments of their own.
3. **Method C.** Regular statistics lectures with no other special requirements.

The difficulties inherent in conducting a meaningful experiment of this type are immense. Yet many such experiments have been conducted by researchers in education; e.g., experiments have been conducted to compare the "new math" curriculum materials with traditional materials.

We will suppose that we have located 12 classes of students for which the instructors have agreed to participate in our experiment, and we are considering whether to use a **completely randomized** design or a **randomized block** design. A completely randomized design would consist of assigning method A to some classes chosen at random from the 12, method B to some classes chosen at random, and method C to the other classes. A randomized block design, on the other hand, would consist of grouping (blocking) the classes into four groups (blocks) of classes and, within each block, assigning the three methods of teaching to the classes at random. In terms explained in Lecture 28, the **methods of teaching** are the **treatments** and the **classes** of students are the **experimental units**. Let us give explicit descriptions of the two designs.

Completely Randomized Design. We have n experimental units and t treatments. Treatment 1 is assigned to n_1 units chosen randomly, treatment 2 to n_2 units chosen randomly, etc. The number of replications ($n_1, n_2, \ldots$) may or may not be equal.

Randomized Block Design. We have bt experimental units grouped into b blocks of t units each and t treatments. Within each block, the treatments are randomly assigned to the experimental units.

30.3 We will now suppose that we have decided to use a completely randomized design and that we have obtained average class scores for each of the 12 classes, as shown in Table 30.1 under the columns headed A, B, and C.

Table 30.1 Data from Completely Randomized Experiment

	A	A^2	B	B^2	C	C^2
	79	6,241	83	6,889	67	4,489
	81	6,561	86	7,396	72	5,184
	86	7,396	91	8,281	74	5,476
	77	5,929	84	7,056	68	4,624
Total	323	26,127	344	29,622	281	19,773

As a **model** for the analysis that follows, we will regard the data arising from three methods as being random samples from three different normal populations with means μ_A, μ_B, and μ_C with a common variance σ^2. The analysis of variance that we give in Table 30.2 is the same as was illustrated in Lecture 28 for a one-way classification, although we will obtain the sums of squares in slightly different fashion.

The **correction factor** (CF) is given by (grand total)$^2/n$. That is,

$$\text{CF} = (948)^2/12 = 74{,}892$$

The total sum of squares is obtained by the sum of squares of every observation minus the correction factor. The **between groups** and **within groups** sums of squares are obtained as shown.

$$\text{Total SS} = 75{,}522 - 74{,}892$$
$$= 630$$
$$\text{Between SS} = \frac{(323)^2}{4} + \frac{(344)^2}{4} + \frac{(281)^2}{4} - \text{CF}$$
$$= 75{,}406.5 - 74{,}892$$
$$= 514.5$$
$$\text{Within SS} = \text{total SS} - \text{between SS}$$
$$= 115.5$$

Because the **between** sum of squares is between groups created by different treatments we will call it the **treatment sum of squares**. We call the **within** sum of squares the **error sum of squares** because the within mean square is an estimate of experimental error variance, which we described in Lecture 28.

Table 30.2 Completely Randomized Analysis of Variance

Source	df	SS	MS	F
Total	11	630		
Treatments	2	514.5	257.250	20.046
Error	9	115.5	12.833	

To test the hypothesis that there are no differences among the treatment effects, we use the F ratio.

$$F = \frac{\text{treatment MS}}{\text{error MS}}$$

$$= 20.046$$

Next we calculate the probability of a larger F. Using Table A.7 with 2 degrees of freedom for the numerator and 9 degrees of freedom for the denominator, we have

$$\text{SL} = P(F > 20.046)$$

$$\doteq .001$$

Using the **significance level** as an **index of credibility** for the hypothesis of equal treatment effects, we would likely conclude that the treatment effects are different. In fact, the sample mean for treatment B is the largest of the three and we would be inclined to accept method B as the most effective way of teaching statistics.

30.4 If, in fact, we have some reason for grouping experimental units (classes in the current example) together as blocks we should consider doing so and then use a **randomized block design**. For example, we might have the situation that the 12 classes are taught by four instructors, each instructor teaching three classes. This groups the classes into four blocks with three classes each and, within each block, we would randomly assign treatments to classes. The average class scores might be as in Table 30.3.

The analysis of variance for the randomized block design was worked out by Fisher in the early 1920s and was given in a footnote to a paper of "Student's" in 1923 [1]. It partitions the total variation into variation

Table 30.3 Data for Randomized Block Design

	Treatments						
Blocks	A	A^2	B	B^2	C	C^2	Total
1	74	5,476	79	6,241	63	3,969	216
2	76	5,776	83	6,889	75	5,625	234
3	84	7,056	89	7,921	67	4,489	240
4	79	6,241	94	8,836	76	5,776	249
Total	313	24,549	345	29,887	281	19,859	939

due to **blocks**, **treatments**, and **experimental error**. The computation of these sums of squares proceeds as follows.

$$\text{CF} = \frac{(939)^2}{12}$$

$$= 73{,}476.75$$

$$\text{Total SS} = 74{,}295 - \text{CF}$$

$$= 818.25$$

$$\text{Blocks SS} = \frac{(216)^2}{3} + \frac{(234)^2}{3} + \frac{(240)^2}{3} + \frac{(249)^2}{3} - \text{CF}$$

$$= 73{,}671 - \text{CF}$$

$$= 194.25$$

$$\text{Treatments SS} = \frac{(313)^2}{4} + \frac{(345)^2}{4} + \frac{(281)^2}{4} - \text{CF}$$

$$= 73{,}988.75 - \text{CF}$$

$$= 512.00$$

$$\text{Error SS} = \text{total SS} - \text{block SS} - \text{treatment SS}$$

$$= 112.00$$

We have avoided giving algebraic formulas, hoping to convey the method of calculation by example. One characteristic of analysis of variance calculations is that when a number is squared, it is almost always divided by the number of basic observations going into it. An inspection of the foregoing calculations will show this happens several times.

Finally, we assemble our calculations, including an F value and a significance level, as in Table 30.4. Again, because of the small value of the significance level, we would be inclined to feel that the teaching methods do have different effects.

Table 30.4 Analysis of Variance—Randomized Block

Source	df	SS	MS	F	SL
Total	11	818.25			
Blocks	3	194.25			
Treatments	2	512.00	256.00	13.71	.005
Error	6	112.00	18.67		

30.5 In the last three lectures we have indicated the power of statistics in experimental science. We have restricted ourselves to elementary examples and have avoided mathematical notation as much as possible. This will afford the students a glimpse of the statistical design of experiments.

SUMMARY. The **completely randomized** design and the **randomized block** designs are extensions of the two-group and paired designs, respectively, to experiments with more than two treatments. With the completely randomized design, treatment 1 is assigned to some units chosen randomly, treatment 2 to some units chosen randomly, etc. With the randomized block design, the treatments are assigned at random to the units within **each** block. With the completely randomized design the sources of variation are **total**, **treatments**, and **error**. With the randomized block design the sources of variation are **total**, **blocks**, **treatments**, and error.

With either design the F ratio for treatments is formed by dividing the **treatment mean square** by the **error mean square**. The **observed significance level** is the probability of getting a larger F by chance when there are no differences among the treatment effects.

REFERENCES

1. Cochran, W. G., and G. M. Cox. (1957). *Experimental Designs*, 2nd Edition, New York: Wiley.
2. Snedecor, G. W., and W. G. Cochran. (1967). *Statistical Methods*, 6th Edition, Ames: Iowa State University Press.
3. "Student." (1923). "On Testing Varieties of Cereals," *Biometrika*, *15*, 271–293.

EXERCISES

1. Verify the analysis of variance in Table 30.2 and the resulting significance level.
2. Verify the analysis of variance in Table 30.4 and the resulting significance level.
3. A director of single-student housing was concerned with the effect on student scholarship of housing policies and conducted a randomized block experiment. Twenty-four students, having given their permis-

sion to random dormitory assignment, were blocked into four blocks of six students each. The blocking was done on the basis of previous grade point averages—the students within any block having similar scholastic records. The treatments were six dormitories with distinctly different rules and regulations. Within each block of students, the students were randomly assigned to these six dormitories. At the end of the year, the grade point averages were as follows.

	Blocks			
Treatments	1	2	3	4
1	3.45	3.68	2.25	2.68
2	2.98	3.35	2.10	2.45
3	2.93	3.17	2.32	2.36
4	3.23	3.52	2.26	2.39
5	3.17	3.95	2.47	2.78
6	3.01	3.78	2.34	2.64

Calculate an analysis of variance table similar to Table 30.4 and calculate a significance level to test the hypothesis of equal treatment effects. Given these data, do you think dormitory assignment affects grades? What is your opinion about the quality of the experiment?

4. A distinguished author has coauthored books with four different people. It is of interest to study whether the writing style is markedly different in the four books. One small study concentrated on the frequency of the use of prepositions. Passages were selected from each of the books, and the number of prepositions in 20-line sections were recorded.

Book			
1	2	3	4
28	29	26	39
31	33	24	27
17	33	22	35
25	35	19	34
26	24	23	28
22	28	25	34
24		29	33
		30	

a. Calculate the analysis of variance and F statistic to test equality of the mean number of prepositions.
b. Calculate the observed significance level.
c. Identify the treatments and experimental units.

5. A grocer is experimenting with different display schemes in the soft drink sections of several stores. He designs a randomized block design to compare the effects of three different schemes on the total weekly sales volume (dollars) in soft drinks. In each store each display scheme is tried for one week, the order being decided at random in each store.

	Scheme		
Store	*A*	*B*	*C*
1	9250	8670	7675
2	1125	1075	1250
3	2420	2780	2525
4	3465	3615	3625
5	4213	3860	3855

a. Calculate the analysis of variance and F statistic to test equality of mean sales volume for the three schemes.
b. Calculate the observed significance level.
c. Identify the treatments and experimental units.

6. A student-faculty committee investigating the effectiveness of student evaluation forms designed a randomized block experiment. Three different versions of the evaluation form were designed. On form A all statements were positively worded. On form B all statements were negatively worded. Form C used some negative statements and some positive statements. Three departments from each of four different colleges participated in the study. Within each college the forms were randomly assigned to departments. The average departmental faculty ratings are given here.

	College			
Form	1	2	3	4
A	2.65	2.58	2.76	2.76
B	2.91	2.74	2.87	3.01
C	2.72	2.61	2.53	2.82

a. Calculate the analysis of variance and F statistic to test equality of mean ratings for the three forms.

b. Calculate the observed significance level.
c. Identify the treatments and experimental units.

7. Students from a psychology class volunteered to participate in a reading experiment. The time required to read a word list was recorded. Three types of word lists were prepared, distinguished from each other by the number of words repeated at least twice. The word lists were assigned at random to students, and the reading times (in seconds) were as shown.

Word Lists		
1	2	3
10.1	14.2	15.2
10.2	15.1	12.9
11.5	16.2	14.3
15.3	13.7	14.8
14.6	13.0	15.0
13.1	12.6	14.7
13.9	14.3	13.1

a. Calculate the analysis of variance and F statistic to test for equality of mean reading times.
b. Calculate the observed significance level.
c. Identify the treatments and experimental units.

8. The trainer for a university football team designs a randomized block design to compare the effect of different breakfasts on running speed of the players. The breakfasts are cereal, ham and eggs, toast and rolls, and pancakes. The blocks consist of four players each from the offensive backs, interior linemen, pass receivers, and defensive backs. An hour after breakfast the players run a 100-yard dash, and the times are recorded. Calculate the analysis of variance and test the hypothesis that there is no difference among breakfasts on the speed of the players.

	Blocks			
	1	2	3	4
Cereal	9.8	10.2	10.1	10.0
Ham and eggs	10.2	10.8	10.2	10.5
Toast and rolls	10.0	10.2	10.4	10.2
Pancakes	9.7	10.8	10.2	10.3

9. The staff of a university newspaper is experimenting with different formats. A completely randomized experiment is designed to compare three different formats. Volunteer students in a journalism class are given one of three formats and, after reading the news stories, take a reading comprehension test. This gives the following statistics.

Format	n	$\overline{X}$	s^2
1	10	90	4.37
2	10	86	3.76
3	11	83	4.21

Perform the analysis of variance, calculate the F statistic, and determine the significance level for the hypothesis that the mean reading comprehension is the same for all formats.

IMPORTANT CONFLICTS OF IDEAS

THE BAYES FAMILY VAULT. Grave of Thomas Bayes in Bunhill Fields, burial ground for many notable nonconformists. Reprinted with permission of Stephen Stigler.

31

THE GREAT BAYESIAN CONTROVERSY

31.1 About $\frac{1}{2}$ mile north of Gresham College, where Karl Pearson gave his lectures on probability and statistics, lies Bunhill Fields. Across City Road from the chapel and home of John Wesley, Bunhill Fields, burial ground for 120,000 souls, is the famous cemetery of the nonconformists, disused since 1852. Here are the graves of John Bunyan, Daniel Defoe, Isaac Watts, and Susannah Wesley, among many other famous people. Here also are the graves of Richard Price and Thomas Bayes.

Readers of current statistics textbooks and journals may search in vain for the names of Fisher and Pearson but may see repeated references to Bayes. Beginning students might well conclude that Bayes, not Pearson, was the founder of modern statistics! In fact, Bayes made almost no contribution to statistics. Why, then, the repeated references to Bayes? The answer seems to lie in a paper by Bayes [1] that has become the focal point for the mode of reasoning called Bayesian statistics. This paper, published in 1763, was submitted after Bayes' death by his friend, Richard Price, who was known for the development of life insurance principles. The paper contained what has become known as **Bayes' theorem**, which consists of nothing more than an elaboration of the rules for conditional probability given in Lecture 8. To understand why so much attention has been given to Bayes' theorem, it is necessary to remind ourselves of ways of thinking about probability.

31.2 In Lecture 7 we discussed the interpretation of probability and emphasized that there are essentially two ways of conceiving of probability: (1) as a **physical property** inherent to a physical system, and (2) as a **measure of belief** in the truth of some statement. Hacking [4] also identifies these two major conceptions of probability. From the time of Pearson's first lectures in probability and statistics until the late 1950s, the overwhelming majority of statisticians held the **first** view of probability with the probability of an event being the relative frequency with which the event occurred or might occur. During this time, Bayes' theorem was used only in situations where probability could be interpreted within the frequency framework. Its use in statistics was almost nonexistent. Rethinking on the meaning of probability has led to a tide of papers in statistics in which probability is regarded as a measure of belief. The recent surge of publications in Bayesian statistics started in 1960. Virtually all of about 175 references in Lindley's 1970 bibliography on Bayesian statistics [5] are to papers written in the 1960s.

Since 1960 the case **for** and **against** Bayesian statistics has been stated and argued with all of the passion and heat that marked the Pearson-Fisher quarrel. Always the argument seems to hinge on how we interpret probability. Let us now try to illustrate why this is crucial for the use of Bayes' theorem.

31.3 Take a simple urn problem of the type so useful in statistics. We are presented with an urn known to be **either** urn I **or** urn II. The contents of the urns are as shown:

Urn I. 4 red balls and 1 white ball
Urn II. 1 red ball and 4 white balls

We draw one ball at random from the urn presented to us and obtain a red ball. What is the probability that we have been presented with urn I, given that we obtained a red ball? Using the rules given in Lecture 8, we would proceed as follows.

$$P(\text{I} \mid \text{red}) = \frac{P(\text{I and red})}{P(\text{red})}$$
$$= \frac{P(\text{red} \mid \text{I})P(\text{I})}{P(\text{red})}$$

Now a red ball can occur in two mutually exclusive ways: with urn I or with urn II. Therefore,

$$P(\text{red}) = P(\text{I and red}) + P(\text{II and red})$$
$$= P(\text{red} \mid \text{I})P(\text{I}) + P(\text{red} \mid \text{II})P(\text{II})$$

so, finally, we have

$$P(\mathrm{I}\,|\,\mathrm{red}) = \frac{P(\mathrm{red}\,|\,\mathrm{I})P(\mathrm{I})}{P(\mathrm{red}\,|\,\mathrm{I})P(\mathrm{I}) + P(\mathrm{red}\,|\,\mathrm{II})P(\mathrm{II})}$$

$$= \frac{(4/5)P(\mathrm{I})}{(4/5)P(\mathrm{I}) + (1/5)P(\mathrm{II})}$$

In order to get a numerical answer we need $P(\mathrm{I})$ and $P(\mathrm{II})$, but these have **not** been given in the statement of the problem. The **frequentist** point of view is that there is **no** way to proceed further in the calculation of the desired probability. The view may be that the state of the urn (either urn I or urn II) is completely fixed and, since there is no randomness, there is no way of discussing $P(\mathrm{I})$ or $P(\mathrm{II})$. On the other hand, even if we know that the choice of urn was made by a random process, we do not know the probabilities associated with urn I and urn II and can proceed no further.

The **Bayesian** point of view goes something like the following. Probability is a measure of our belief, and we can **always** express our degree of belief in terms of probability. Complete ignorance about whether the urn was I or II would be expressed with probabilities of 1/2 for both. Using these probabilities, we would obtain

$$P(\mathrm{I}\,|\,\mathrm{red}) = \frac{(4/5)(1/2)}{(4/5)(1/2) + (1/5)(1/2)}$$

$$= 4/5$$

On the other hand, suppose we strongly believed that the turn was urn II. We might use $P(\mathrm{I}) = 1/10$ and $P(\mathrm{II}) = 9/10$. Then we would have

$$P(\mathrm{I}\,|\,\mathrm{red}) = \frac{(4/5)(1/10)}{(4/5)(1/10) + (1/5)(9/10)}$$

$$= \frac{4}{4+9}$$

$$= 4/13$$

Continuing to think of probability as a measure of belief, look at these calculations. The occurrence of a red ball should increase the probability of urn I. This it has done in both cases; in the first case from $P(\mathrm{I}) = 1/2$ to $P(\mathrm{I}\,|\,\mathrm{red}) = 4/5$ and in the second from $P(\mathrm{I}) = 1/10$ to $P(\mathrm{I}\,|\,\mathrm{red}) = 4/13$.

Incidentally, the idea of using equal probabilities to represent complete ignorance was used by Bayes in the 1763 paper and is called the **Bayes postulate**. It is also known among other names, such as **Laplace's principle of insufficient reason.**

31.4 We will now use the simple urn example as a base for describing Bayesian statistics. We can think of the urn number as a **parameter**, and our **data** consist of the color of the ball drawn. Before we draw the ball and observe its color, we have probabilities about the parameter value [P(I) and P(II)]. The probabilities are called **prior probabilities** because they exist before or prior to our collecting data. After we collect our data, it seems reasonable to calculate new probabilities, which are called **posterior probabilities**, or after-the-fact probabilities. These posterior probabilities are, using Bayes' theorem, the **conditional probabilities** of the parameter value given the data.

Now consider the problem of making inferences about the mean μ of a normal population on the basis of a sample from that normal population. Statistics based on a **frequency interpretation** of probability treats μ as a fixed parameter for which **no** probability statements can be made, either before or after the sample. Inferences about μ are based on estimates of μ, calculation of significance levels, consonance intervals, etc. However, the **Bayesian** approach is to use **prior** probabilities for μ and Bayes' theorem to obtain **posterior** probabilities for μ.

31.5 It is still not clear how the controversy is going to be resolved. Prior to 1960 virtually all statistics avoided the Bayesian approach. Fisher had steadfastly avoided the use of Bayes' theorem and, in his last book [3], strongly rejected it once more. Although Pearson had on occasion used Bayes' theorem, he had, in the main, avoided it. Neyman and Pearson had decided against its use in their landmark paper on testing hypotheses. However, Bayesian statistics has attracted many followers since 1960, and further development may enhance its usefulness. The case for Bayesian statistics is forcefully stated by Lindley [5]. For a presentation of arguments against Bayesian statistics, refer to the paper by Berkson [2].

SUMMARY. Bayes wrote a paper that was published after his death by his friend Richard Price. This paper, published in 1763, contained a theorem that has become known as **Bayes' theorem**. It allows the calculation of one conditional probability given another conditional probability and certain marginal probabilities. The usefulness of the theorem hinges on the **interpretation of probability**.

If a probability allows two possible values, θ_1, and θ_2, for a parameter θ,

$$P(\theta_1 \mid \text{data}) = \frac{P(\text{data} \mid \theta_1)P(\theta_1)}{P(\text{data} \mid \theta_1)P(\theta_1) + P(\text{data} \mid \theta_2)P(\theta_2)}$$

In order to use the formula, $P(\theta_1)$ and $P(\theta_2)$ must be known. The Bayesian view of probability is as a **measure of belief**, so $P(\theta_1)$ and $P(\theta_2)$ are considered as measures of belief **prior** to collecting the sample. After collecting the data, these probabilities are modified by calculating the conditional probabilities $P(\theta_1 \mid \text{data})$ and $P(\theta_2 \mid \text{data})$. These latter probabilities are called **posterior** probabilities.

REFERENCES

1. Bayes, Thomas. (1763). "An Essay Towards Solving a Problem in the Doctrine of Chances," *Philosophical Transactions of the Royal Society*, *53*, 370–418.
2. Berkson, Joseph. (1977). "My Encounter with Neo-Bayesianism," *International Statistical Review*, *45*, 1–8.
3. Fisher, R. A. (1956). *Statistical Methods and Scientific Inference*, Edinburgh: Oliver and Boyd.
4. Hacking, Ian. (1974). *The Emergence of Probability*, London: Cambridge University Press.
5. Lindley, D. V. (1970). *Bayesian Statistics, A Review*, Regional Conference Series in Applied Mathematics, 2. Philadelphia: SIAM.

UNIVERSITY COLLEGE LONDON. The main college building viewed from the south. Courtesy University College London, England.

32

THE FISHER-PEARSON CONTROVERSY

32.1 Karl Pearson was born in London in 1857. Before his death in 1936 he was to author more than **600** articles and books covering subjects from the history of art to theoretical statistics. He must be regarded as the founder of modern statistics. A summary of his life is given by his son, E. S. Pearson [4]. For other articles see Walker [8], Stouffer [7], Wilks [9], and Haldane [2].

What was to become a long association with University College London began when he was sent to University College School in 1866. He was withdrawn by his parents for health reasons and placed under a private tutor at Hitchin, about 30 miles north of London, in 1873. A year later he went to Cambridge and worked under another tutor, as was customary. In 1875 he was awarded a scholarship at King's College, Cambridge, and took his degree with mathematical honors in 1879.

Supported by a King's College Fellowship, the next several years were for Pearson devoted to an incredible program of work, study, and travel. He studied and wrote in Germany, and in London he studied law and was called to the bar in 1881. He taught some classes at King's College, London, and at University College London. In 1884 he was appointed to the Chair of Applied Mathematics and Mechanics at University College London. During the next several years he worked hard at teaching mathematics to engineering students and at lecturing and writing on widely ranging subjects.

32.2 In London there is still a building at the corner of Gresham and Basinghall streets with a worn sign that says Gresham College. Sir Thomas Gresham (1519–1579), English financier and member of Queen

Elizabeth's privy council, was the founder of the Royal Exchange. He invested the rents from 100 shops in the Royal Exchange for the purpose of establishing a college in which seven professors were to lecture on "divynitye, astronomy, musick, geometry, law, physicke, and rethoricke." From 1579 until 1768 the college was housed in Gresham's old home, a medieval manor house. Thereafter, lectures were delivered at the Royal Exchange until 1838. A new college building was erected at the present location in 1842; the present building, housing the Graduate Business Centre of the City of London University, dates from 1913. In the year that R. A. Fisher was born, Pearson was appointed as the nineteenth lecturer in geometry at Gresham College on December 15, 1890. He held this appointment concurrently with his chair at University College until he resigned the Gresham College lectureship on June 12, 1894. At Gresham College he gave a series of popular lectures on probability and statistics; syllabuses of these lectures are included in the volume by E. S. Pearson [4]. The modern age of statistics had begun. In these lectures in 1893 he first used the expression **standard deviation** and began referring to the probability **law of errors** as the **normal distribution**.

Following the Gresham College lectures, Pearson gave at University College from 1894 to 1895 and 1895 to 1896 the **first university courses** on the **theory of statistics**. An outline of these courses is included by E. S. Pearson [4].

Karl Pearson is called the founder of **biometrics**, and his laboratories at University College are often referenced. Modern visitors to University College may search in vain for traces of the Biometrics Laboratory, although the Galton Laboratory still exists in fact and name in the Department of Human Genetics and Biometry. Pearson himself traced the beginnings of the Biometrics Laboratory to 1895. With Francis Galton, Pearson founded the journal *Biometrika*; the first volume appeared in 1901. In 1907 he accepted supervision of the Galton Laboratory, founded 3 years earlier by Francis Galton. Upon the death of Francis Galton in 1911, the Galton Chair of Eugenics was established at the bequest of Galton, and Pearson gave up his chair of applied mathematics to accept the Galton chair and to establish a new department, the Department of Applied Statistics. In 1912 Fisher published his first paper.

In 1911 money was given by Sir Herbert H. Bartlett to construct the building on Gower Street to house the new department. The building was completed in 1914 but, because of World War I, was not fully occupied until 1920. The statistics program at University College is still housed there.

In 1925, at 68 years of age and with a brilliant career already behind him, Pearson launched another major activity with the founding of the *Annals of Eugenics*, a journal devoted to the study of human genetics. In more recent years the name has been changed to the *Annals of Human Genetics.*

In 1933, with much regret, Pearson resigned his chair, and his department was split. Fisher was appointed to the Galton chair to head the Department of Eugenics, and Pearson's son, E. S. Pearson, was named to head the new Department of Applied Statistics.

Pearson continued to work in his beloved profession until he died in 1936.

32.3 An account of the life of Fisher is given by Kendall [3] and Yates and Mather [10]. The entire issue of *Biometrics* (June 1964) is devoted to Fisher, and the article by Pearson [6] provides a glimpse of Fisher at University College. Ronald A. Fisher was born in London in 1890. During the years that the foundations were being laid for biometry and modern statistics at University College, Fisher was receiving a good education at Stanmore Park School and Harrow School. He entered Cambridge University, where he specialized in mathematics, in 1909. His first scientific paper was published in 1912, the same year that he graduated from Cambridge, and 3 years later his famous paper giving the probability distribution of the **correlation coefficient *r*** appeared in *Biometrika.* Although Fisher subsequently submitted a few papers to *Biometrika* for consideration, he never published another paper in that journal. In 1919 he was offered a post at University College, but he accepted instead the job at Rothamsted (see Lecture 25).

Until his death in 1962, he produced a flood of books and papers on many aspects of applied as well as theoretical statistics. The most famous statistics books are *Statistical Methods for Research Workers* (1925) and *The Design of Experiments* (1935). His work brought him fame throughout the statistical world, and prizes and honors were heaped on him. He was knighted in 1952.

We have already mentioned that Fisher succeeded Pearson in 1933 in the Galton chair at University College. In 1943 he returned to Cambridge to accept the Arthur Balfour Chair of Genetics. In 1957 he retired, but a few years later he accepted a research fellowship in Australia in the Division of Mathematical Statistics of the Commonwealth in Adelaide. He died there following an operation in 1962.

32.4 For many years the statistical world was enlivened by the fierce dispute that developed between Pearson and Fisher. There may be little point in trying to determine exactly how the dispute materialized. However, the fact of its existence is an important part of the story of statistics, and any students who read the works of Fisher and Pearson will become aware of this fact. Egon Pearson [5] gives some interesting clues to its origins that are contained in correspondence among Fisher, "Student," and Pearson.

The disagreements between Fisher and Pearson involved more than personalities; they involved different philosophies of statistics. The Pearsonian viewpoint, particularly as modified and developed by Pearson and Neyman, has lead to a view of statistics as a **decision-making** science, while the statistical heirs of Fisher have come to think of statistics more as a science of **data analysis** and **formation of opinion**. We will discuss these viewpoints further in Lecture 33.

SUMMARY. Karl Pearson (1857–1937) and R. A. Fisher (1890–1962) were contemporaries and fellow pioneers in statistics during parts of their lives. Pearson had already established his fame and reputation before Fisher began his career. In 1884 Pearson was appointed to the Chair of Applied Mathematics and Mechanics at University College London. In 1901 he edited the first volume of *Biometrika*, and in 1911 he established the Department of Applied Statistics at University College London. In **1912** Fisher graduated from Cambridge and published his **first paper**, and in **1915** he published his first statistical paper, on the distribution of the **sample correlation coefficient**.

Between 1915 and 1919 serious differences, never to be breached, arose between Pearson and Fisher. From 1919 to 1933 Fisher was at Rothamsted establishing the foundations of **experimental statistics**. During this same period, Pearson's department prospered at University College. Pearson retired in 1933 and died 3 years later. In 1933 Fisher became chairman of the Department of Eugenics at University College.

Both men made enormous contributions to statistics. Yet their philosophies of science and of statistics were different. To some extent these different philosophies in statistics persist to this day.

REFERENCES

1. Box, Joan Fisher. (1978). *R. A. Fisher. The Life of a Scientist*, New York: Wiley.

2. Featherstone, Ernest. (1952). *Sir Thomas Gresham and His Trusts*, London: Blades, East, and Blades.
3. Haldane, J. B. S. (1957). "Centenary Lecture: Karl Pearson, 1857–1957," *Biometrika, 44*, 303–313.
4. Kendall, M. G. (1963). "Ronald Aylmer Fisher, 1890–1962," *Biometrika, 50*, 1–5.
5. Pearson, E. S. (1938). *Karl Pearson, An Appreciation of Some Aspects of His Life and Work*, Cambridge, England: Cambridge University Press.
6. Pearson, E. S. (1968). "Some Early Correspondence Between W. S. Gosset, R A. Fisher, and Karl Pearson, with Notes and Comments," *Biometrika, 55*, 445–457.
7. Pearson, E. S. (1974). "Memories of the Impact of Fisher's Work in the 1920's," *International Statistical Review, 42*, 5–8.
8. Stouffer, Samuel A. (1958). "Karl Pearson—An Appreciation on the 100th Anniversary of His Birth," *Journal of the American Statistical Association, 53*, 23–27.
9. Walker, Helen M. (1958). "The Contributions of Karl Pearson," *Journal of the American Statistical Association, 53*, 11–22.
10. Wilks, S. S. (1941). "Karl Pearson: Founder of the Science of Statistics," *The Scientific Monthly, 53*, 249–253.
11. Yates, F., and D. Mather. (1963). Foreword to *Collected Works of R. A. Fisher*, reprinted from *Biographical Memoirs of Fellows of the Royal Society of London, 9*, 91–120.

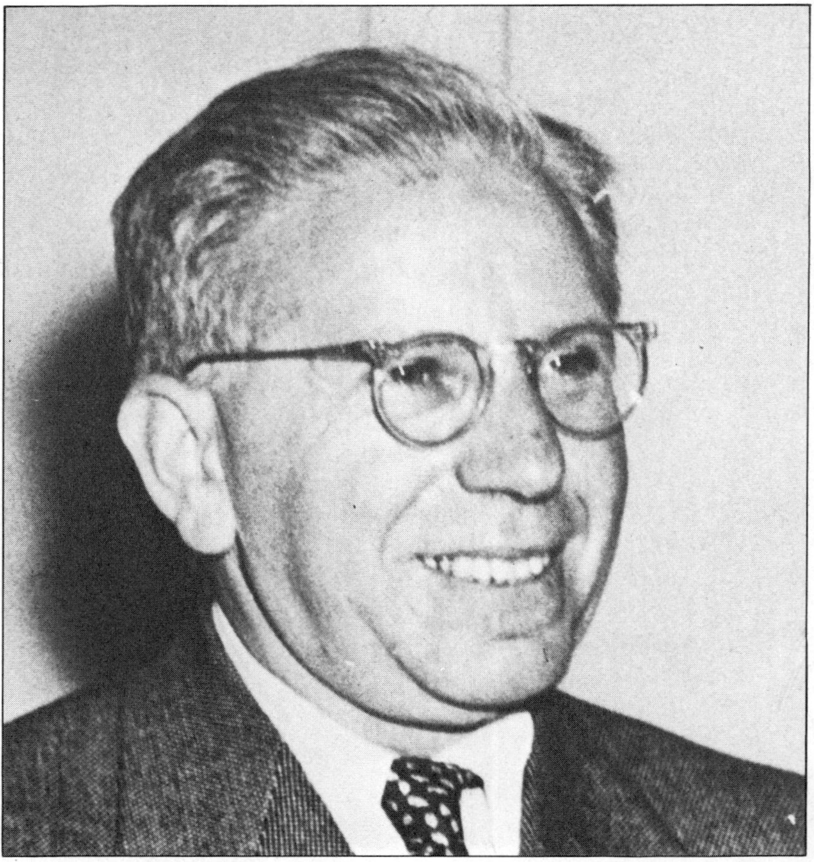

ABRAHAM WALD. 1902–1950. Smith Collection, Rare Book and Manuscript Library, Columbia University.

33

INFERENCE VERSUS DECISION

33.1 In 1950 the book by Wald entitled *Statistical Decision Functions* [3] introduced a new era in statistical thinking. Wald's theory places the statistician or the user of statistics in a **decision-making** situation. Data are to be analyzed and used to reach one of a possible set of decisions.

This formulation, which is consistent with the Neyman-Pearson theory of **testing hypothesis** (accept H_0 or H_A), has been accepted by many statisticians as the fundamental way to describe all statistics. This leads to the description of statistics as the science of decision making under uncertainty. Wald apparently visualized his work as a unifying theory of statistics. In the preface he states:

*A major advance beyond previous results is the treatment of the design of experimentation as a part of the general decision problem.**

Fisher reacted strongly to the suggestion that Wald's decision theory included the design of experiments. He maintained that although decision theory might be suitable for industry and technology, it was not appropriate for scientific work. In his last book [1, p. 100] he stated:

. . . it would still be true that the Natural Sciences can only be successfully conducted by responsible and independent thinkers applying their minds and their imaginations to the detailed interpretation of verifiable observations. The idea that this responsibility can be delegated to a vast computer programmed with Decision Functions belongs to the phantasy of circles rather remote from scientific research.†

* Reproduced by permission of the publishers, John Wiley & Sons, Inc., from *Statistical Decision Functions* by Abraham Wald, 1950.

† Reproduced by permission of the publishers, Hafner Press, MacMillan Publishing Co., Inc., New York, from *Statistical Methods and Scientific Inference* by R. A. Fisher. Copyright © 1973 by the University of Adelaide.

The process of learning from statistical analysis as opposed to reaching a decision has come to be known as **statistical inference**. The relationship of statistical inference to statistical theory is exposited clearly by Menges [2], who refers to the English Fisher school when discussing inference. This is probably appropriate, since much of the philosophy of inference statistics can be traced to the ideas of Fisher. On the other hand, the concept of statistics as a decision science is clearly more directly related to the Neyman-Pearson theory.

33.2 In earlier lectures I introduced **estimation**, **tests**, and **intervals**, which are fundamental analyses in statistical inference. Let us now consider some simple decision problems and give some elements of modern decision theory.

We are going to buy a car that is required for our work. We expect to keep the car for 2 years, during which time we expect to drive 40,000 miles. We have narrowed the choice of cars to two: car *A* costing \$5000 and averaging 20 miles per gallon and car *B* costing \$6700 and averaging 40 miles per gallon. Which car should we choose if we are concerned only with the total of the cost of the car and the gasoline cost for the 2 years? There is no way to solve the problem without including the price of the gasoline. Consider working the problem with three possible average prices of gasoline: \$1 per gallon, \$2 per gallon, and \$3 per gallon. The total costs for the 2-year period are given in Table 33.1.

We now have the information on which to base a decision. If we think the average price of gasoline will be about \$1, we should choose car *A*. If we think it will be about \$2, we should choose car *B*. Without knowing what the price of gasoline will be, what decision should we make?

One rule, called the **minimax rule**, is to consider the **worst** possible state of affairs for each possible decision and to choose the decision for which the worst state of affairs is **best**. That is, it considers the maximum

Table 33.1 Total Cost—Car Decision Problem

	Prices		
	\$1	\$2	\$3
Car *A*: \$5,000 car using 2,000 gallons	\$7,000	\$9,000	\$11,000
Car *B*: \$6,700 car using 1,000 gallons	\$7,700	\$8,700	\$9,700

total cost for car A and for car B and chooses the car giving the minimum of the maximum costs (minimax rule). With car A, the maximum cost is \$11,000 dollars; with car B, the maximum cost is \$9700 dollars. So the **minimax** decision is car B. As Menges [2] states, this rule is not generally recommended by statisticians because it is too pessimistic.

Another rule, called the **Bayes rule**, requires specification of probabilities for the prices of gasoline. Clearly, in this instance these probabilities are measures of belief. Suppose we feel rather strongly that the price of gasoline will remain low, so that we specify probabilities of $\frac{3}{4}$, $\frac{1}{8}$, and $\frac{1}{8}$ for \$1, \$2, and \$3 gasoline. Next we calculate the average cost for car A and car B. The Bayes rule is to choose the car for which the average cost is the smallest. We can easily verify that

$$\text{Ave(cost for car } A) = (\tfrac{3}{4})(7000) + (\tfrac{1}{8})(9000) + (\tfrac{1}{8})(11{,}000)$$

$$= 7750$$

$$\text{Ave(cost for car } B) = 8075$$

Then car A is the **Bayes decision** for the probabilities specified. Other choices of probabilities will lead to car B as the Bayes decision.

33.3 The previous example illustrates many of the features of a decision problem. There is a **parameter space** (unknown price of gasoline), a **decision space** (car A or car B), and a **cost function**, or **loss function** (Table 33.1). However, this example does not include data that give us some partial information about the parameter. Consider a second example for which the formulation includes **data**. We will give a simplified version of a problem that might occur in the sampling inspection of a manufacturing firm.

When manufacturing lots are of low quality but pass inspection, it is felt that the loss in customer **goodwill** is considerable. Lots that do not pass inspection are reworked—the cost of good lots being slightly less than poorer ones. The possible situations are portrayed in Table 33.2.

Table 33.2 Losses

	Fraction Defective		
	.05	.10	.15
Pass	0	1000	1000
Rework	300	500	500

A sample of two items is taken from each lot, and x = number of defective items observed. Two possible decision rules are being considered.

Rule 1. Pass the lot if $x = 0$.

Rule 2. Pass the lot if $x = 0$ or 1.

The probabilities of passing and reworking for both rules for each fraction defective are given in Table 33.3. By multiplying the probabilities by the losses and summing, we can obtain, for each fraction defective and for each decision rule, the **average loss (risk)**. For example, the **risk** for Rule 1 when the fraction defective is .10 is given by

$$\text{Risk} = (.90)^2(1000) + [1 - (.90)^2](500)$$
$$= 905$$

Proceeding in this way, we obtain the full set of risks given in Table 33.4.

Rule 1 is the **minimax** rule. Let us determine the Bayes rule if our probabilities of .05, .10, and .15 are .90, .05, and .05, respectively. Using these probabilities, we obtain the expected risks.

Rule 1. 114.6375
Rule 2. 99.8625

So, for the probabilities given, the **Bayes rule** is Rule 2.

Table 33.3 Probabilities of Pass and Rework

	Fraction Defective					
	.05		.10		.15	
	Rule 1	Rule 2	Rule 1	Rule 2	Rule 1	Rule 2
Pass	$(.95)^2$	$1-(.05)^2$	$(.90)^2$	$1-(.10)^2$	$(.85)^2$	$1-(.15)^2$
Rework	$1-(.95)^2$	$(.05)^2$	$1-(.90)^2$	$(.10)^2$	$1-(.85)^2$	$(.15)^2$

Table 33.4 Risks

	Fraction Defective		
	.05	.10	.15
Rule 1	29.25	905	861.25
Rule 2	0.75	995	988.75

33.4 It seems that the decision theory just introduced would have all of the necessary components to deal with **parameter estimation**, **tests**, and **intervals**. In fact, it seems that this is true only in a formal sense in scientific work. Researchers do not generally know what losses (or rewards) are likely to result from their work nor do they have a clearly defined set of decisions or probabilities for the possible parameter values. For work requiring business decisions, on the other hand, decision theory seems admirably suited.

SUMMARY. Wald's book, **Statistical Decision Functions**, was an important landmark in the development of statistics as a **decision-making** science. Fisher resisted attempts to systematize statistics to such an extent, apparently preferring to regard statistics more in terms of data analysis and formation of opinion. Learning from data instead of making decisions is called **statistical inference**.

Statistical decision theory is formulated as follows. A person has several decisions open to him or her in the face of several unknown **states of nature**. For each **action-state** pair there is a loss (possibly negative) to the decision maker. One of the possible decisions is made on the basis of **sample data**. The expected value of the loss over possible samples is called the **risk**. The decision maker wishes to minimize the risk in some way. Two rules are discussed: the **minimax rule** and the **Bayes rule**. The minimax rule is to take the choice for which the maximum risk is smallest. The Bayes rule is to take the choice for which the expected risk (with respect to a probability distribution on the states of nature) is smallest.

Statistical decision theory has not succeeded in unifying statistical methods, although many statistical problems may be properly thought of as decision problems.

REFERENCES

1. Fisher, R. A. (1965). *Statistical Methods and Scientific Inference*, Edinburgh: Oliver & Boyd.
2. Menges, G. (1973). "Inference and Decision," in *Selecta Statistica Canadiana*, Vol. 1, New York: Wiley.
3. Wald, A. (1950). *Statistical Decision Functions*, New York: Wiley.

EXERCISES

1. Given the cost information in Section 33.2, verify that the total costs for cars A and B are shown in Table 33.1.

2. What are the Bayes decisions in Section 33.2 for the following sets of probabilities?

	\$1	\$2	\$3
a.	$\frac{1}{4}$	$\frac{1}{2}$	$\frac{1}{4}$
b.	$\frac{1}{4}$	$\frac{1}{4}$	$\frac{1}{2}$
c.	$\frac{1}{8}$	$\frac{1}{8}$	$\frac{3}{4}$

3. Verify the risks in Table 33.4.

4. What are the Bayes decisions in Section 33.3 for the following sets of probabilities?

a.	.95	.025	.025
b.	.90	.10	0
c.	.80	.10	.10

5. Given the loss function in Table 33.2, suppose that a sample of two items is to be taken from each lot. Determine the risk table for the following rules.

   Rule 1. Pass the lot if $x = 0$.
   Rule 2. Pass the lot if $x = 0$ or 1.
   Rule 3. Pass the lot if $x = 0$, 1, or 2.

   With this risk table, what are the Bayes rules for the sets of probabilities in Exercise 4?

6. A collector is considering purchase of a painting, reputed to be by a well-known artist. The painting is priced at \$5000. If genuine, it is worth \$10,000; if false, it is worthless. Furthermore, purchase of a fake painting or failure to buy a genuine one will damage her reputation. The loss table is as follows.

	Genuine (dollars)	Fake (dollars)
Buy	−5000	6000
Not	3000	0

She goes to an appraiser who can detect genuine articles with probability .95 and fake articles with probability .70. Determine the risk table for the following three rules:

Rule 1. Buy with probability $\frac{1}{2}$.
Rule 2. Buy if appraiser says genuine.
Rule 3. Do not buy.

7. Determine the minimax rule for the risk function in Exercise 6.
8. Determine the Bayes rule for the risk function in Exercise 6 for the following sets of probabilities:
   a. $\frac{1}{4}$ $\frac{3}{4}$.
   b. $\frac{1}{2}$ $\frac{1}{2}$.
   c. $\frac{3}{4}$ $\frac{1}{4}$.

MANY DIMENSIONS

CHI-SQUARE. The independence of smoking or nonsmoking and the incidence of disease is often tested using chi-square. Copyright © Gerri Marshall 1980/Design Conceptions.

34

CHI-SQUARE

34.1 In the course of these lectures a few of the well-known statistics were introduced, such as the **correlation coefficient**, **"Student's"** t, and **Snedecor's** F. Although we have previously made reference to the **chi-square** statistic, we have deferred a fuller description until now. The chi-square distribution is a **skewed** distribution, indexed by one parameter, commonly called the **degrees of freedom**, and is tabulated in Table A.8. Statistics having the chi-square distribution arise in a variety of ways; in this lecture we will illustrate some of the many uses of chi-square.

34.2 In 1900 Pearson [3] published a paper on the uses of chi-square, illustrating its use in testing the hypothesis that a sample of data has come from a **completely specified** distribution. For example, we might have a set of measurement data for which the histogram looks somewhat like a Poisson distribution with a specified parameter. We would like to have a test to measure the extent to which the data support the normal hypothesis. Pearson provided such a test. The steps involved are as follows.

1. Arbitrarily set **intervals** covering the range of the data.
2. **Count** the number of observations in each of the intervals. Denote these counts by $O_1, O_2, \ldots, O_k$, where k is the number of intervals.
3. **Calculate** the **probability** for each interval specified by the hypothesized distribution.
4. Calculate the **expected number** of observations, E, for each interval by multiplying the sample size, n, times the calculated probabilities: $E_1 = nP_1$, $E_2 = nP_2$, etc.

5. Calculate the **statistic** $\Sigma(O_i - E_i)^2/E_i$. Pearson showed that the distribution of this statistic is well approximated by the chi-square distribution tabulated in Table A.8. The statistic is therefore widely known as the chi-square statistic and is denoted by χ^2.
6. Calculate the **significance level** from Table A.8 by finding the probability of a larger value than that observed. Pearson used $k - 1$ (the number of intervals minus one) as the degrees of freedom, regardless of whether the parameters were known or were estimated from the data. However, Fisher [1, 2] showed that a better approximation is given by decreasing the degrees of freedom by one for each parameter estimated from the data.

34.3 We will illustrate the **goodness-of-fit** test described in the previous section for two data sets.

In Table 5.5 we presented the **horsekick** data of Bortkiewicz. In order to test the hypothesis that the data come from a **Poisson** distribution, we estimate the parameter by the sample mean $\bar{x} = 0.61$. Using Poisson tables, we can then calculate probabilities and expected numbers by multiplying these probabilities by $n = 200$. These quantities are given in Table 34.1. From the last two lines of Table 34.1, we calculate the χ^2 statistic.

$$\chi^2 = \frac{(O - E)^2}{E}$$

$$= 0.7920$$

From Table A.8 we see that the significance level is quite large (using $k - 1 - 1 = 3$ degrees of freedom), so the hypothesis of a Poisson distribution is acceptable.

Consider the hypothesis that the data of Table 34.2 came from a **normal** distribution.

Table 34.1 Calculation of χ^2—Poisson

x	0	1	2	3	4
O_i	109	65	22	3	1
P_i	0.544	0.331	0.101	0.021	0.003
E_i	108.8	66.2	20.2	4.2	0.6
$O_i - E_i$	0.2	−1.2	1.8	−1.2	0.4

Table 34.2 Distribution of Brain Weight

Brain Weight (grams)	Frequency
Under 1100	0
1100–1150	1
1150–1200	10
1200–1250	21
1250–1300	44
1300–1350	53
1350–1400	86
1400–1450	72
1450–1500	60
1500–1550	28
1550–1600	25
1600–1650	12
1650–1700	3
1700–1750	1
1750 and over	0

Source. Reproduced from *Biometrika, 4,* Pearl, Biometrical studies on man I. Variation and correlation in brain-weight. Copyright © 1905 by the Biometrika Trustees. Reprinted by permission from the Biometrika Trustees.

From the data we estimate the mean and standard deviation by

$$\bar{x} = 1400.4807$$

and

$$s = 107.30420$$

Using these estimated values, we can obtain probabilities from Table A.3 and multiply by the sample size $n = 416$ to obtain expected values. These quantities are shown in Table 34.3.

From Table 34.3 we can calculate our chi-square statistic to be $\chi^2 = 11.6945$, and from Table A.8 we see that the probability of a larger value with $k - 1 - 2 = 15 - 1 - 2 = 12$ df is about .5. Therefore we conclude that the hypothesis of a normal distribution is reasonable.

34.4 Another use of chi-square is in the testing for **independence** of row and column classifications in **two-way** tables. Again we give an example. In a large statistics class attended by graduate students as well as under-

Table 34.3 Calculation of χ^2—Normal

Interval	O_i	P_i	E_i
Below 1100	0	.0026	1.0816
1100–1150	1	.0073	3.0368
1150–1200	10	.0208	8.6528
1200–1250	21	.0501	20.8416
1250–1300	44	.0928	38.6048
1300–1350	53	.1456	60.5696
1350–1400	86	.1808	75.2128
1400–1450	72	.1772	73.7152
1450–1500	60	.1466	60.9856
1500–1550	28	.0954	39.6864
1550–1600	25	.0494	20.5504
1600–1650	12	.0215	8.9440
1650–1700	3	.0073	3.0368
1700–1750	1	.0020	.8320
Above 1750	0	.0006	.2496

graduate students, grading was a major problem for the instructor. Some undergraduates felt that the grading policy gave preferential treatment to the graduate students. Consider the data of Table 34.4. The number of each letter grade assigned is recorded for four different groups of students. It is of interest to ask if the **distribution of grades** is different for

Table 34.4 Test for Independence

		A	*B*	*C*	*D*	Total
Junior	*O*	4	5	7	6	22
	E	6.3871	7.8065	5.3226	2.4839	
Senior	*O*	8	6	6	4	24
	E	6.9677	8.5161	5.8065	2.7097	
M.S.	*O*	16	14	11	4	45
	E	13.0645	15.9677	10.8871	5.0806	
Ph.D.	*O*	8	19	6	0	33
	E	9.5806	11.7097	7.9839	3.7258	
Total		36	44	30	14	124

the different levels of students. Suppose that the probability of receiving an A is independent of the student classification. Then

$$P(\text{junior and } A) = P(\text{junior})P(A)$$

Under the hypothesis of independence, this probability is estimated by (36/124)(22/124), and the expected number is estimated by multiplying by 124. Therefore the expected number for the junior A cell is given by $(36)(22)/(124) = 6.3871$. We can obtain the other expected numbers in the table similarly. Then we calculate the chi-square statistic to be $\chi^2 = 19.0684$.

The number of degrees of freedom for a table with r rows and c columns is simply $(r - 1)(c - 1)$. This follows from our previous rule for degrees of freedom. Under the hypothesis of independence, we need to estimate only the row probabilities and column probabilities. Since the row probabilities must total 1, we need to estimate only $r - 1$ of these and, similarly, $c - 1$ column probabilities. Therefore the degrees of freedom are

$$rc - 1 - (r - 1) - (c - 1) = (r - 1)(c - 1)$$

For the current example, the degrees of freedom are 9. From Table A.8, the probability of a chi-square with 9 degrees of freedom exceeding 19.0684 is about .025. Because of this relatively small value, we would tend to accept the idea that the distribution of grades is **not** independent of the student classification.

34.5 In recent years other **goodness-of-fit** statistics have gained preference over **chi-square**, yet it continues to be widely used and some knowledge of it seems almost mandatory.

SUMMARY. Pearson provided a chi-square **goodness-of-fit** test to test the conformity of a set of data to a hypothesized distribution. A frequency distribution is constructed for a sample of data. For each of the classes or intervals, the **expected** or theoretical **frequencies** are calculated for the specified distribution. Then the chi-square statistic is calculated using the formula $\Sigma(O - E)^2/E$, where O denotes **observed frequencies** and E denotes **expected frequencies**. The **degrees of freedom** (as modified by Fisher) are $n -$(number of classes)$-$(number of parameters estimated).

Another use of the chi-square statistics is to test **independence** of the two criteria of classification in a two-way table of frequency counts. In this case, the observed frequency for any cell is (row total) (column total)/(table total).

The degrees of freedom for the chi-square statistic are $(r - 1)(c - 1)$, where r = number of rows and c = number of columns.

REFERENCES

1. Fisher, R. A. (1922). "On the Interpretation of χ^2 from Contingency Tables and the Calculation of *P*," *Journal of the Royal Society*, *85*, 87–94.
2. Fisher, R. A. (1924). "The Conditions Under Which χ^2 Measures the Discrepancy Between Observation and Hypothesis," *Journal of the Royal Society*, *87*, 442–250.
3. Pearson, Karl. (1900). "On the Criterion that a Given System of Deviations from the Probable in the Case of a Correlated System of Variables is Such that It Can Be Reasonably Supposed to Have Arisen From Random Sampling," *Philosophical Magazine*, *6* (2), 157–176.

EXERCISES

1. Verify the chi-square calculations in Table 34.1 and the resulting significance level for the horsekick data.
2. Examine the goodness of fit of the Poisson distribution to "Student's" yeast cell data (Table 12.2) by calculating the chi-square statistic and evaluating the significance level.
3. Verify that the formula $k - 1 -$ (number of parameters estimated) does result in $(r - 1)(c - 1)$ for the degrees of freedom as in Section 34.4.
4. Verify the value for $\bar{x}$ and s in Section 34.3 and verify the calculation of χ^2 in Table 34.3.
5. In the case of a continuous distribution the choice of the grouping interval is arbitrary and has some effect on the chi-square test. Recalculate the value of chi-square for the data of Table 34.2 using intervals 1100–1200, 1200–1300, etc. Are you led to the same conclusions concerning normality of the data?

6. Examine the goodness of fit of the normal distribution to the data of Table 5.2.

7. Sociologists often study the effect of parental attitudes on attitudes of the children. A survey was made of high school students in a city and they were asked about their participation in church activities. Church attendance was reported as follows.

	Parents		
Children	Once a Week	Once a Month	Seldom
Once a week	74	67	27
Once a month	11	14	33
Seldom	5	10	17

Calculate the chi-square to test for independence of parental church attendance from church attendance of the children. Determine the significance level.

8. High school football players were asked questions to determine their plans to attend college after graduation. The researcher was interested in the possible relationship between the strength of their high school program and the strength of the football program at the college of their choice. The season record for students' high school football teams and their chosen college football teams were as follows.

	High School Team Win Record (percent)			
College Team Win Record (percent)	0–25	25–50	50–75	75–100
0–50	25	26	11	2
50–100	21	23	12	17

Calculate the chi-square statistic and the significance level to test for independence.

9. A public opinion research company surveyed attitudes of people concerning a proposed cut in state taxes. Calculate the chi-square

statistic to test for independence and the significance level for the following set of data.

	Favor	Oppose
State employees	198	217
Not state employees	476	109

10. The people working in an office building were asked about the comfort of their office (too cold, about right, too warm) in October, January, and April. Calculate the chi-square test for independence and significance level.

	October	January	April
Too cold	231	168	212
About right	66	89	79
Too warm	148	113	136

11. People were asked for their opinion about salaries of public school teachers (too high, too low, about right). Some people questioned were teachers, some had teachers in their immediate families, and others were not closely related to teachers. Calculate the chi-square statistic to test for independence and the significance level for the following set of data.

	Teacher	Related	Not Related
Too high	2	102	624
About right	37	378	578
Too low	316	314	309

CHARLES EDWARD SPEARMAN. 1863–1945. Keystone Press Agency.

35

FACTOR ANALYSIS

35.1 Since its early beginnings with contributions from Galton [1888] and Pearson [1901], **factor analysis** has been developed and pushed to its present state primarily by psychologists. In 1904 Spearman [4] presented the idea of measuring **general intelligence**. Suppose that we have aptitude scores for an individual in mathematics, science, and art. Then, according to the general intelligence theory, these three scores would result from a combination of a **general** intelligence score for the individual and **specific** aptitude scores for the three areas.

35.2 In order to carry the matter further we must discuss some preliminary concepts. Recall that a **standardized** variable is one that has mean zero and variance unity. A variable is standardized by subtracting its mean and then dividing by its standard deviation. In psychology as well as in other social sciences, much use is made of standardized variables, and we will generally assume in this discussion of factor analysis that the observation variables have been standardized. We are trying to express these standardized variables in terms of a basic, underlying, but unmeasurable general factor.

35.3 Suppose the **factor structure** is known and the standardized aptitude scores Z_1, Z_2, and Z_3 for mathematics, science, and art are given by linear combinations of a standardized **general** intelligence score, F, and **specific** aptitude scores e_1, e_2, and e_3 as follows.

$$Z_1 \text{ (mathematics)} = .9F + e_1$$
$$Z_2 \text{ (science)} = .8F + e_2$$
$$Z_3 \text{ (art)} = .1F + e_3$$

We also suppose that F and the es are **uncorrelated**.

The coefficients of F are called **factor-loading coefficients**—the idea being that they describe the way Z_1, Z_2, and Z_3 "load onto" the underlying factor F. Thus the mathematics and science scores are more strongly related to the factor F than is the art score. We might therefore associate the factor F with mathematics and science aptitudes. This idea can be made a little more precise because the factor-loading coefficients give the correlation coefficients between the Zs and F. So the correlation coefficient between Z_1 and F is .9, between Z_2 and F .8, and between Z_3 and F .1.

The squares of the factor-loading coefficients are called **communalities** and are also directly interpretable. They give the fractions of the variances of the Zs accounted for by the common factor F. The remaining portions of the variances are called **specificities**. For the example at hand we have the following data.

	Total Variance	Communality	Specificity
Z_1	1.0	.81	.19
Z_2	1.0	.64	.36
Z_3	1.0	.01	.99

We could say, for example, that 99% of the variance of the aptitude score for art is due **specifically** to art and only 1% is due to the **common** factor involving mathematics, science, and art.

35.4 The simple models suggested by Spearman's work were soon generalized to include more than one "general intelligence" factor. Suppose that we had two uncorrelated general scores F_1 and F_2 and that the standardized scores Z_1, Z_2, and Z_3 were given as follows.

$$Z_1 = .9F_1 + .1F_2 + e_1$$

$$Z_2 = .8F_1 + .2F_2 + e_2$$

$$Z_3 = .1F_1 + .8F_2 + e_3$$

The interpretation of the factor-loading coefficients as correlation coefficients carries forward directly and, in fact, provides some interpretation for the factors. The general factor F_1 has correlation .9, .8, and .1 with Z_1, Z_2, and Z_3, while the general factor F_2 has correlations of .1, .2, and .8. This suggests that F_1 may be a **scientific** factor while F_2 may be an **aesthetic** factor.

It is also possible to calculate the **communalities** and specificities from the factor-loading coefficients. This requires forming the matrix product AA' from the matrix, A, of **factor-loading coefficients**. For this example,

$$AA' = \begin{bmatrix} .9 & .1 \\ .8 & .2 \\ .1 & .8 \end{bmatrix} \begin{bmatrix} .9 & .8 & .1 \\ .1 & .2 & .8 \end{bmatrix} = \begin{bmatrix} .82 & .74 & .17 \\ .74 & .68 & .24 \\ .17 & .24 & .65 \end{bmatrix}$$

The **diagonal** elements of AA' give the **communalities**. So, for this two-factor model, we have:

	Total Variance	Communality	Specificity
Z_1	1.0	.82	.18
Z_2	1.0	.68	.32
Z_3	1.0	.65	.35

35.5 We have developed the discussion in the previous section as though we knew the factor-loading coefficients (the coefficients of the Fs) and as though we could measure the Fs and the es. In fact, the usual situation is that we do not know the factor-loading coefficients and we cannot measure the Fs and the es; we can measure only the Zs. From measurements of the Zs we must try to estimate the coefficients, the Fs, and the es. This can be done in many ways, and a casual survey of the factor analysis literature will show that many methods are available. Estimation procedures are readily available in packages of statistical computer programs.

The use of factor analysis has been successful in a number of cases, particularly when the factors, F, are known or suspected in advance. On the other hand, factor analysis has been employed to analyze many sets of data, hoping to discover underlying factors, without preconceived ideas of what they might be. Successful applications in these circumstances are somewhat rare.

35.6 Another way to approach factor analysis is the use of ***n*-dimensional** geometry. Science fiction writers invent fantasy worlds by introducing a fourth dimension, and the idea is quite intriguing. In a very real sense we actually live in a world of more than three dimensions, but it is not the world of science fiction.

To illustrate the idea of multidimensional space, consider a population of people. For each person we can obtain physical measurements such as height, weight, waist size, etc. We can also obtain measurements such as IQ scores and aptitude test scores. Suppose, in addition, that we obtain quantitative measures of characteristics such as courage, tenacity, and honesty. Then each person can be thought of as a long list of scores. Now imagine that we have an axis for length, one for weight, and one for courage. We can plot a person's scores in n-dimensional space, and each person can be thought of as a **point** in n-dimensional space. A population of people is a **cluster** of points, like an n-dimensional balloon. A population of people similar in nature would be a tightly packed cluster, while a population of people with diverse natures would form a loose cluster. Presumably the stereotypic, well-adjusted individuals are points in the center of the cluster. The maladjusted and abnormal as well as the exceptional are points on the fringes of the cluster.

If we have obtained three scores on each of 50 people we can conceive of our total set of data as 50 points in three-dimensional space. More generally, we will have obtained p scores on each of n people and can think of our data as ***n*** **points** in ***p*****-dimensional space**. Our n points form some sort of cluster in p-dimensions. Typically, we hope that it is an **elliptical cluster** (a p-dimensional cigar), and we try to find the **principal axes** of the cluster. Most of the variation lies along the first principal axis, the next largest amount of variation along the second principal axis, etc. Of course, we appeal to our geometric intuition, but the methodology is algebraic. The method is known as **principal components**, and the first paper on the subject was by Pearson (1901).

SUMMARY. Spearman conceived of the idea that scores on tests are combinations of a general intelligence factor and a factor **specific** to mathematics, art, music, etc. This led to a **factor model** in which a **standardized** test score is expressed as a constant times a standard, unmeasurable **factor** score plus a standard, unmeasurable **specific** score. The constant is called the **factor-loading coefficient** and is also the correlation between the measurable test score and the unmeasurable factor score. The variance of the test score is the sum of two components: a part common to all test scores called the **communality** and a part specific to a particular test score called the **specificity**.

The single-factor model was later generalized to allow for **several factors**. The computation without a computer program remains onerous, but many programs are available.

REFERENCES

1. Galton, Francis (1888). "Co-relations and Their Measurements, Chiefly from Anthropometric Data," *Proceedings of the Royal Society*, *45*, 135–140.
2. Harman, Harry H. (1976). *Modern Factor Analysis*, 3rd Edition, Chicago: The University of Chicago Press.
3. Pearson, Karl. (1901). "On Lines and Planes of Closest Fit to Systems of Points in Space," *Philosophical Magazine*, *2–6*, 559–572.
4. Spearman, C. (1904). "General Intelligence, Objectively Determined and Measured," *American Journal of Psychology*, *15*, 201–293.

EXERCISES

1. F_1 and F_2 are uncorrelated factor scores (standardized variables), and standardized scores Z_1, Z_2, and Z_3 are given by

$$Z_1 = .8F_1 + .1F_2 + e_1$$
$$Z_2 = .1F_1 + .7F_2 + e_2$$
$$Z_3 = .2F_1 + .6F_2 + e_3$$

   a. What are the correlation coefficients between the Zs and the Fs?
   b. What is the basic interpretation of the Fs in terms of the Zs?
   c. Calculate the communalities and specificities.

2. A student researcher conducted a survey of campers during his summer vacation. He visited with other campers at different campsites and obtained responses to a questionnaire concerning their attitude toward vacation activities. Upon returning to the campus in the fall, he analyzed his data using a factor analysis program from the computing center and identified two factors. The coefficients are as follows.

Z	F_1	F_2
Hiking	.56	−.08
Fishing	.47	.03
Boating	.39	.10
Swimming	.45	.06
Ranger talks	.07	.56
Guided tours	.22	.34
Camping chores	.14	.41
Campfire	−.09	.53

a. How would you interpret the factors identified?
b. Calculate the communalities and specificities.

3. Most computer centers have packages of statistical procedures available, and a factor analysis procedure is usually included as one of the programs. The users manual for a factor analysis program gives the following factor-loading coefficients for a problem involving three measurements on each of five subjects.

	Factor		
Subject	1	2	3
1	.3	.7	.1
2	.7	.1	.2
3	.8	.1	.1
4	.2	.8	.1
5	.1	.6	.1

Calculate the communalities and specificities for the three factors. Which subjects typify factor 1? Which subjects typify factor 2?

4. An examination for a class of 20 students had five questions, two intended to measure retention of the lecture material and three intended to measure the ability of the students to apply the material. The factor-loading coefficients for the two important factors identified follow.

	Factor	
Question	1	2
1	.91	.02
2	.15	.78
3	−.88	.11
4	.79	−.12
5	.09	.86

Which questions were intended to measure retention? Which questions were intended to measure ability to apply the material?

5. The students in a large undergraduate class volunteer to take standard examinations in college algebra, accounting, English

composition, American history, and chemistry. The factor-loading coefficients for the two major factors are given here. How would you interpret the two factors?

Exam	Factor 1	Factor 2
College algebra	.78	.52
Accounting	.81	.59
English composition	.76	−.61
American history	.69	−.66
Chemistry	.72	.68

P. C. MAHALANOBIS. 1893–1972. Indian Statistical Institute, Calcutta.

36

DISCRIMINANT FUNCTION ANALYSIS

36.1 There is a strong desire on the part of many to predict periods of upswing as well as plunges in the stock market. Attempts to do so usually make use of some function of previous stock market prices. Similarly, we would like to be able to identify the potential dropouts and the potential persisters among entering college freshmen by examining their high school records. As another example, a physician would like to be able to discriminate between patients likely to have heart attacks and those who will not by looking at their medical records. Some success has been achieved on these and many other discrimination problems, but there is still a desire to make the discrimination sharper. In this lecture we introduce **statistical discriminant functions**, which have been used to attempt the sort of discrimination referred to in these examples.

The general approach is to use data from two or more known populations where the population identity of each individual is known. One then devises a discriminant function, which does as good a job as possible of sorting the individuals in the sample into the correct populations. If the **classification** is extremely good, one will have the confidence of correctly classifying an individual known to be from one of the populations, the specific population unknown.

36.2 We will now develop some of the ideas of discriminant function analysis by use of an example. Consider the data in Table 36.1. IQ scores and high school grade point averages have been obtained for 24 college freshmen. Twelve withdrew from college during their freshman year, and the other 12 continued to their sophomore year.

Given these data, we set out to find a **discriminant function**. When we consider X_1 (IQ), our hopes begin to fade. There is a slight difference

Table 36.1 College Freshmen

	Dropouts		Persisters	
	X_1 (IQ)	X_2 (HS Average)	X_1 (IQ)	X_2 (HS Average)
	109	2.50	107	2.54
	108	2.68	105	2.65
	113	2.54	106	2.91
	115	2.86	111	2.70
	116	2.79	112	2.86
	114	3.03	113	3.24
	119	3.00	114	3.11
	121	3.25	118	3.10
	122	3.15	118	3.26
	122	3.37	119	3.25
	124	3.35	123	3.27
	128	3.62	121	3.49
Total	1411	36.14	1367	36.38
Mean	117.58	30.12	113.92	3.03

between the means (117.58 and 113.92) in the two samples, but a t test quickly reveals that the difference is not significant and due to chance. Even if the difference were significant, the average IQ for the persisters is lower than for the dropouts, and we would be inclined to be distrustful of the results.

We encounter a similar experience when we examine X_2 (high school grade point average). The difference in sample means is very slight and is of no significance, as judged by a t test.

Our hopes dashed, we might read further about discriminant functions to learn that we are encouraged to examine the variables **jointly**. This we have **not** yet done. We have examined X_1 and X_2 **separately**. We decide to make another effort.

36.3 One of the simplest ways of analyzing two variables is to plot the data in the form of a scattergram. This we have done in Figure 36.1. Once we have made this plot, our hopes of discriminating between the dropouts and persisters on the basis of X_1 and X_2 increase. We can see from the scattergram that the data may have come from two overlapping populations. We might try separating the data as much as possible into

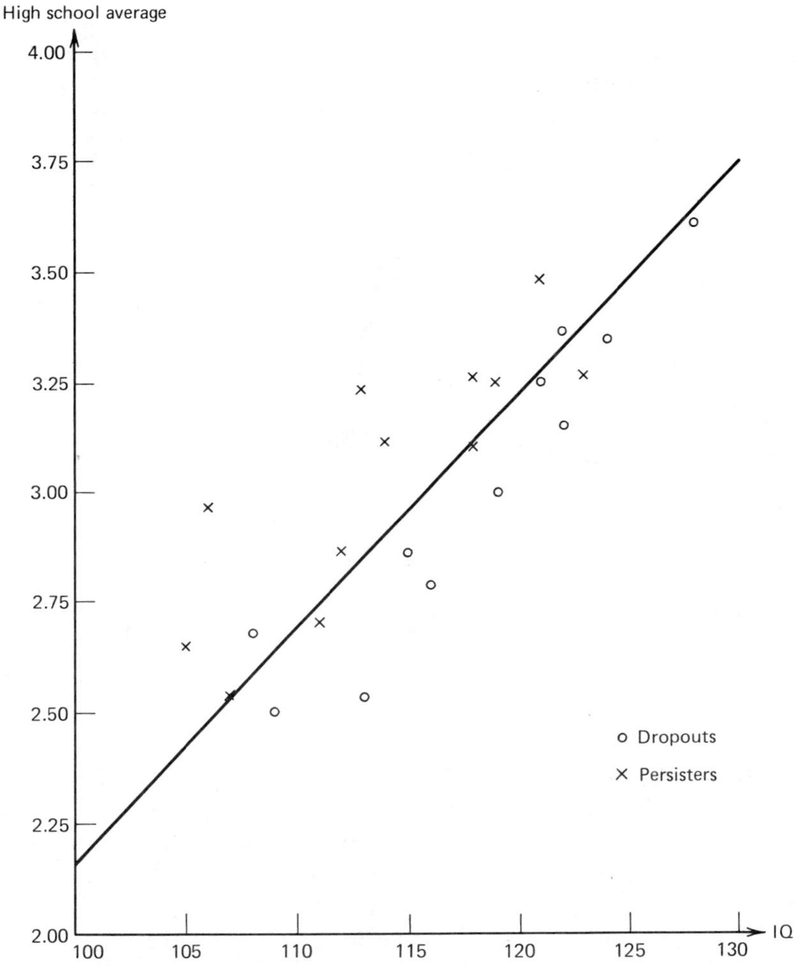

Figure 36.1 Scattergram: College Freshmen.

two clusters by a straight line. This is analogous to distinguishing between two long mountain ridges by looking up the valley between them. As long as we try to look across the ridges, we have difficulty in distinguishing between them, as when we look only at X_1 (or X_2) we are unable to distinguish between dropouts and persisters.

In Figure 36.1 we have drawn a straight line. We have done this merely by looking at the points and without any theoretical reason for choice of line. In fact, considerable theoretical research has been done on the choice of line, which is not discussed in this lecture. Note that the line passes through the points (100, 2.15) and (130, 3.75). We thus ascertain the slope to be (3.75 − 2.15)/(130 − 100) = .053. We also determine the X_2-intercept to be 2.15 − (100)(.053) = −3.15. Therefore the equation of the straight line that we have chosen is

$$X_2 = -3.15 + .053X_1$$

or, equivalently,

$$.053X_1 - X_2 = 3.15$$

We can also determine that

$$.053X_1 - X_2 < 3.15 \qquad \text{for points above the line}$$

and

$$.053X_1 - X_2 > 3.15 \qquad \text{for points below the line}$$

We have thus arrived at a **discriminant function** $.053X_1 - X_2$ and a **classification rule**. Classify an individual as a **dropout** if $.053X_1 - X_2 > 3.15$. Otherwise, call that student a **persister**.

36.4 We now see how well this classification rule works for the data of Table 36.1. Of course, we can see from Figure 36.1 that the straight line leaves three persisters as well as three dropouts on the wrong side of the line. However, it is worthwhile to evaluate the discriminant function chosen for each of the data points. These values, with the resulting classifications, are given in Table 36.2.

From Table 36.2 we see that our discriminant function performs very well. We might summarize this performance as in Table 36.3.

36.5 Given more variables, the simple method given in this lecture will not work. However, we can still use it to aid our intuition. We can imagine two p-dimensional **overlapping clusters** of points. We try to separate the two clusters by passing a plane between them. Of course, this is actually done algebraically, and the plane is a linear combination of our variables. Computer programs are available that will calculate **discriminant functions**. A full understanding of what is taking place requires

Table 36.2 Discriminant Analysis

Dropouts		Persisters	
Function	Classification	Function	Classification
3.277	Dropout	3.131	Persister
3.044	Persister	2.915	Persister
3.449	Dropout	2.708	Persister
3.235	Dropout	3.183	Dropout
3.358	Dropout	3.076	Persister
3.012	Persister	2.749	Persister
3.307	Dropout	2.932	Persister
3.163	Dropout	3.154	Dropout
3.316	Dropout	2.994	Persister
3.096	Persister	3.057	Persister
3.222	Dropout	3.249	Dropout
3.164	Dropout	2.923	Persister

Table 36.3 Performance of Discriminant Function

	Classification	
Group	Number Correct	Number Incorrect
Dropout	9	3
Persister	9	3

a knowledge of statistics and p-dimensional geometry, but discriminant function analysis can be useful to one familiar only with the simple ideas in this lecture.

36.6 Discriminant functions have been used widely by paleontologists. Given a bone fragment (jawbone, elbow, etc.) judged to be quite old by all dating techniques, we are interested in whether or not the fragment is human. McHenry [5] gives a detailed discussion of the matter.

36.7 In 1936 both Mahalanobis [4] and Fisher [2] published important papers contributing to the development of discriminant functions.

SUMMARY. Frequently sample data are available from **two populations** that are known to be **different**. It is desired to find a function of the data that will have rather different values for the two samples. If this function shows the samples to have come from different populations, the hope is that it would **classify** by population **future** observations.

The basic idea of **linear discriminant functions** can be obtained by plotting observations from two **overlapping** populations. When each variable is regarded separately, it may be impossible to see any separation. However, when two or more variables are considered jointly, it may be that a straight line (or plane) **separates** the populations fairly well. The equation for the line gives a **linear discriminant function**.

REFERENCES

1. Bolch, B. W., and C. J. Huang. (1974). *Multivariate Statistical Methods for Business and Economics*, Englewood Cliffs, N.J.: Prentice-Hall.
2. Fisher, R. A. (1936). "The Use of Multiple Measurements in Taxonomic Problems," *Ann. Eugenics*, *7*, 179–188.
3. Kendall, M. G. (1968). *A Course in Multivariate Analysis*, New York: Hafner.
4. Mahalanobis, P. C. (1936). "On the Generalized Distance in Statistics," *Proc. Nat. Inst. Sci., India*, *12*, 49–55.
5. McHenry, Henry M. (1975). "Fossils and the Mosaic Nature of Human Evolution," *Science*, *190* (4213), 425–431.
6. Morrison, D. F. (1967). *Multivariate Statistical Methods*, New York: McGraw-Hill.

EXERCISES

1. Verify the discriminant function values and classifications given in Table 36.2.
2. The following data on state finances were given by the U.S. Census Bureau for the year 1973. All figures are in millions of dollars.

State	Revenue (per capita)	Debt (per capita)	Expenditure (per capita)
Southeast			
Alabama	478	848	454
Arkansas	449	109	397
Florida	460	1,260	430
Georgia	484	755	471
Kentucky	531	1,816	505
Louisiana	565	1,214	554
Mississippi	529	584	495
North Carolina	495	528	451
South Carolina	498	603	469
Tennessee	422	642	386
Virginia	479	388	464
West Virginia	595	842	582
Northeast			
Connecticut	560	2,580	526
Maine	531	358	526
Massachussetts	582	3,105	602
New Hampshire	408	179	420
New Jersey	456	2,755	443
New York	756	11,801	721
Pennsylvania	532	4,596	530
Rhode Island	587	391	553
Vermont	742	421	754

Source. U.S. Bureau of the Census, Government Finances in 1972–1973.

Plot a scattergram of revenue versus debt and try to obtain a linear discriminant function to discriminate between northeastern states and southeastern states. Also try revenue versus expenditure and debt versus expenditure. Using what seems to be the most successful discriminant function, construct tables similar to Tables 36.2 and 36.3.

3. Do the best job possible of correctly identifying by population the following samples using a rule of the form: If $X_1 < C$ classify in one population; if $X_1 > C$ classify in the other population. Do the same using X_2 and then X_3. Next use large and small values of $2X_1 + X_2 + 3X_3$.

Population 1			Population 2		
X_1	X_2	X_3	X_1	X_2	X_3
2	1	2	−3	2	6
6	−2	0	−1	3	5
−3	2	5	6	2	0
3	2	1	0	2	4
1	2	2	2	5	2
6	2	−1	3	3	2
10	0	−3	10	10	−3
11	8	−6	−3	−3	8
−11	2	9			

APPENDIX

Table A.1 Random Numbers

15 77 01 64 69	69 58 40 81 16	60 20 00 84 22	28 26 46 66 36	86 66 17 34 49
85 40 51 40 10	15 33 94 11 65	57 62 94 04 99	05 57 22 71 77	99 68 12 11 14
47 69 35 90 95	16 17 45 86 29	16 70 48 02 00	59 33 93 28 58	34 32 24 34 07
13 26 87 40 20	40 81 46 08 09	74 99 16 92 99	85 19 01 23 11	74 00 79 41 69
10 55 33 20 47	54 16 86 11 16	59 34 71 55 84	03 48 17 60 13	38 71 23 91 83
05 06 67 26 77	14 85 40 52 68	60 41 94 98 18	62 20 94 03 71	60 26 45 17 92
65 50 89 18 74	42 07 50 15 69	86 97 40 25 88	14 17 73 92 07	93 11 93 45 15
59 68 53 31 55	73 47 16 49 79	69 80 76 16 60	58 53 07 04 53	66 94 94 18 13
31 31 05 36 48	75 16 00 21 11	42 44 84 46 84	83 20 49 17 12	21 93 34 61 16
91 59 46 44 45	49 25 36 12 07	25 90 89 55 25	83 47 17 23 93	99 56 14 39 16
63 59 73 21 67	80 00 25 58 25	72 06 12 86 74	54 79 70 85 88	71 58 21 98 48
89 72 47 46 94	78 56 10 65 97	84 79 42 31 49	94 15 31 13 09	45 43 03 82 81
70 51 21 03 18	50 21 99 49 73	06 99 19 24 96	39 43 10 14 12	94 08 55 54 70
14 15 99 60 44	62 72 38 18 36	63 92 61 55 93	77 66 82 10 91	81 51 67 01 47
92 46 90 39 99	64 08 00 97 27	54 96 63 40 54	34 70 27 48 18	68 59 91 83 32
81 23 17 13 01	37 57 92 16 34	15 80 90 25 64	67 77 29 95 84	80 84 84 87 22
87 54 42 46 56	28 89 02 06 98	59 90 74 13 38	98 66 23 20 23	90 55 31 83 48
74 73 84 98 13	11 48 25 33 39	27 36 08 99 57	60 42 88 68 25	22 89 67 83 16
94 55 14 00 97	32 51 92 47 03	92 33 73 20 21	29 77 37 06 98	64 63 34 31 43
69 21 94 26 20	73 90 70 92 76	49 14 60 34 43	90 51 72 11 07	75 94 19 49 40
82 36 36 89 29	87 70 08 71 98	49 00 89 89 99	29 08 02 72 32	68 16 29 82 19
25 06 22 30 87	87 44 48 90 91	38 53 10 60 29	40 07 58 97 84	09 04 33 56 72
82 37 97 60 92	76 39 17 84 34	67 65 52 89 90	62 97 04 33 81	91 27 56 46 35
83 71 07 22 15	17 55 56 82 62	88 83 86 38 14	63 89 39 81 90	25 62 58 68 87
73 13 79 15 12	18 34 22 24 75	56 47 45 22 81	30 82 38 34 52	57 48 30 34 17
91 28 00 57 30	92 12 38 95 21	15 70 78 50 88	01 07 90 72 77	99 53 04 34 73
33 47 55 62 57	08 21 77 31 05	64 74 04 93 42	20 19 09 71 46	37 32 69 69 89
56 66 25 32 38	64 70 26 27 67	77 40 04 34 63	98 99 89 31 16	12 90 50 28 96
88 40 52 02 29	82 69 34 50 21	74 00 91 27 52	98 72 03 45 65	30 89 71 45 91
87 63 88 23 62	51 07 69 59 02	89 49 14 98 53	41 92 36 07 76	85 37 84 37 47
32 25 21 15 08	82 34 57 57 35	22 03 33 48 84	37 37 29 38 37	89 76 25 09 69
44 61 88 23 13	01 59 47 64 04	99 59 96 20 30	87 31 33 69 45	58 48 00 83 48
94 44 08 67 79	41 61 41 15 60	11 88 83 24 82	24 07 78 61 89	42 58 88 22 16
13 24 40 09 00	65 46 38 61 12	90 62 41 11 59	85 18 42 61 29	88 76 04 21 80
78 27 84 05 99	85 75 67 80 05	57 05 71 70 21	31 99 99 06 96	53 99 25 13 63
42 39 30 02 34	99 46 68 45 15	19 74 15 50 17	44 80 13 86 38	40 45 82 13 44
04 52 43 96 38	13 83 80 72 34	20 84 56 19 49	59 14 85 42 99	71 16 34 33 79
82 85 77 30 16	69 32 46 46 30	84 20 68 72 98	94 62 63 59 44	00 89 06 15 87
38 48 84 88 24	55 46 48 60 06	90 08 83 83 98	40 90 88 25 26	85 74 55 80 85
91 19 05 68 22	58 04 63 21 16	23 38 25 43 32	98 94 65 35 35	16 91 07 12 43
54 81 87 21 31	40 46 17 62 63	99 71 14 12 64	51 68 50 60 78	22 69 51 98 37
65 43 75 12 91	20 36 25 57 92	33 65 95 48 75	00 06 65 25 90	16 29 34 14 43
49 98 71 31 80	59 57 32 43 07	85 06 64 75 27	29 17 06 11 30	68 70 97 87 21
03 98 68 89 39	71 87 32 14 99	42 10 25 37 30	08 27 75 43 97	54 20 69 93 50
56 04 21 34 92	89 81 52 15 12	84 11 12 66 87	47 21 06 86 08	35 39 52 28 09
48 09 36 95 36	20 82 53 32 89	92 68 50 88 17	37 92 02 23 43	63 24 69 80 91
23 97 10 96 57	74 07 95 26 44	93 08 43 30 41	86 45 74 33 78	84 33 38 76 73
43 97 55 45 98	35 69 45 96 80	46 26 39 96 33	60 20 73 30 79	17 19 03 47 28
40 05 08 50 79	89 58 19 86 48	27 98 99 24 08	94 19 15 81 29	82 14 35 88 03
66 97 10 69 02	25 36 43 71 76	00 67 56 12 69	07 89 55 63 31	50 72 20 33 36

Table A.1 *Continued*

15 62 38 72 92	03 76 09 30 75	77 80 04 24 54	67 60 10 79 26	21 60 03 48 14
77 81 15 14 67	55 24 22 20 55	36 93 67 69 37	72 22 43 46 32	56 15 75 25 12
18 87 05 09 96	45 14 72 41 46	12 67 46 72 02	59 06 17 49 12	73 28 23 52 48
08 58 53 63 66	13 07 04 48 71	39 07 46 96 40	20 86 79 11 81	74 11 15 23 17
16 07 79 57 61	42 19 68 15 12	60 21 59 12 07	04 99 88 22 39	75 16 69 13 84
54 13 05 46 17	05 51 24 53 57	46 51 14 39 17	21 39 89 07 35	47 87 44 36 62
95 27 23 17 39	80 24 44 48 93	75 94 77 09 23	48 75 91 69 03	55 51 09 74 47
22 39 44 74 80	25 95 28 63 90	41 19 48 46 72	51 12 97 39 83	35 83 23 17 29
69 95 21 30 11	98 81 38 00 53	41 40 04 16 78	67 29 83 41 18	30 90 44 37 64
75 75 63 97 12	11 57 05 86 52	82 72 47 72 14	37 72 69 75 48	72 21 52 51 81
08 74 79 30 80	70 11 66 79 25	88 01 94 52 31	38 57 98 71 62	12 56 61 01 54
04 88 45 98 60	90 92 74 77 87	40 18 65 87 37	08 68 62 39 52	84 74 90 68 18
97 35 74 05 75	42 13 49 48 38	74 19 06 42 60	20 79 90 81 77	18 51 71 27 27
53 09 93 28 29	80 19 68 30 45	94 49 49 71 21	93 93 71 30 34	52 65 83 40 13
26 36 68 48 09	37 69 26 22 80	23 34 10 45 70	83 51 07 37 44	62 96 74 42 64
49 16 57 15 79	56 63 22 94 28	11 39 69 55 38	53 06 97 20 42	09 14 90 43 48
03 51 79 78 74	75 23 73 75 98	47 85 07 26 02	61 28 01 22 16	14 12 15 67 22
21 88 87 28 48	23 44 03 03 80	53 89 07 87 93	30 17 84 17 74	16 53 31 39 01
56 41 73 33 41	59 16 59 50 98	24 24 87 06 75	99 52 09 88 05	86 25 43 50 94
72 39 19 70 17	01 04 01 22 33	04 84 63 27 65	84 39 45 55 31	95 88 93 90 37
97 28 25 81 49	71 69 22 04 51	56 46 56 15 10	69 59 99 50 29	33 50 16 93 09
18 87 02 72 08	74 52 16 03 82	20 19 66 23 62	37 51 04 89 31	32 19 59 85 57
53 40 11 75 45	13 56 85 31 37	09 17 71 96 79	39 50 79 27 62	71 14 95 53 03
60 49 03 41 56	78 33 77 28 92	21 90 10 62 01	97 06 45 01 19	95 12 24 18 52
09 16 12 75 04	39 69 95 00 48	26 85 28 73 08	66 92 10 66 75	62 61 27 82 57
64 20 19 87 54	88 15 12 54 24	06 99 57 07 28	51 34 54 98 50	70 88 02 86 48
31 28 07 58 77	03 98 26 76 09	10 44 57 61 28	60 29 85 70 79	80 29 19 98 92
80 04 28 47 76	35 73 67 78 28	09 39 88 63 74	41 26 92 42 33	06 80 06 33 84
24 60 22 51 19	34 54 08 24 73	86 72 11 44 69	76 90 81 17 85	57 47 35 16 84
59 16 11 26 29	18 97 78 44 43	58 92 78 70 80	09 65 32 68 26	65 73 90 50 46
58 54 29 98 27	40 51 92 07 13	58 41 59 56 94	16 32 51 42 54	77 37 13 85 19
20 18 34 22 73	57 40 67 17 28	63 57 74 36 18	65 55 25 50 68	35 90 00 03 38
53 90 46 56 19	50 58 33 84 53	14 74 17 40 73	86 11 04 02 04	02 28 49 62 36
97 16 93 94 65	70 95 95 83 20	91 42 57 95 63	00 86 29 02 53	02 27 86 70 95
72 55 71 70 92	04 22 53 19 29	67 29 13 56 70	45 73 45 05 04	32 43 30 93 41
99 19 72 58 35	49 09 26 00 74	26 42 94 52 02	83 31 85 65 66	31 97 67 52 15
48 21 49 72 97	79 19 64 81 82	78 92 51 96 51	28 79 13 20 82	34 81 39 46 86
52 37 68 15 53	22 98 30 16 31	83 24 87 69 29	24 85 44 25 50	75 62 83 95 41
97 50 52 53 52	26 78 21 68 69	57 79 42 40 89	55 81 75 24 52	51 32 79 97 05
36 05 09 18 11	71 01 63 17 60	11 65 19 43 07	44 86 19 58 92	23 71 32 96 19
20 79 70 09 30	81 14 53 80 93	71 94 10 18 14	83 69 76 53 25	27 36 65 65 05
13 07 89 72 08	00 37 75 14 94	83 85 06 72 66	07 47 30 17 11	16 02 63 97 30
94 26 82 37 43	34 23 00 14 50	96 85 41 17 71	69 20 15 98 82	79 69 68 50 31
13 55 88 38 43	75 37 43 83 85	53 74 54 62 99	68 93 74 43 95	06 26 79 78 87
02 44 24 97 71	97 93 12 70 89	42 52 33 24 91	05 87 53 15 77	49 92 83 97 80
34 90 96 63 54	22 84 36 38 99	85 36 25 03 27	49 24 72 10 50	95 14 18 26 64
13 67 06 34 98	04 20 80 12 54	01 18 54 20 76	92 10 47 04 65	54 45 82 42 90
18 75 55 82 66	34 77 27 71 79	67 65 85 92 68	16 43 83 18 74	12 48 68 87 22
91 25 52 57 15	21 54 40 05 50	67 51 66 45 69	84 72 74 32 30	17 70 40 90 24
76 24 00 14 92	14 29 12 17 73	77 46 44 24 30	48 50 36 30 24	93 08 01 39 37

Source. Extracted from Table IX, *Statistical Tables and Formulas* by A. Hald. Copyright © 1952 by John Wiley & Sons, Inc., New York. Reprinted by permission of John Wiley & Sons, Inc.

Table A.2 Binomial Probabilities

A table of $\binom{n}{x} p^x(1-p)^{n-x}$ for $n = 1, 2, \ldots, 10$ and $p = 0.01, 0.05(0.05)0.30, \frac{1}{3}, 0.35(0.05)0.50$, and $p = 0.49$

n	x \ p	.01	.05	.10	.15	.20	.25	.30	$\frac{1}{3}$	.35	.40	.45	.49	.50
2	0	.9801	.9025	.8100	.7225	.6400	.5625	.4900	.4444	.4225	.3600	.3025	.2601	.2500
	1	.0198	.0950	.1800	.2550	.3200	.3750	.4200	.4444	.4550	.4800	.4950	.4998	.5000
	2	.0001	.0025	.0100	.0225	.0400	.0625	.0900	.1111	.1225	.1600	.2025	.2401	.2500
3	0	.9703	.8574	.7290	.6141	.5120	.4219	.3430	.2963	.2746	.2160	.1664	.1327	.1250
	1	.0294	.1354	.2430	.3251	.3840	.4219	.4410	.4444	.4436	.4320	.4084	.3823	.3750
	2	.0003	.0071	.0270	.0574	.0960	.1406	.1890	.2222	.2389	.2880	.3341	.3674	.3750
	3	.0000	.0001	.0010	.0034	.0080	.0156	.0270	.0370	.0429	.0640	.0911	.1176	.1250
4	0	.9606	.8145	.6561	.5220	.4096	.3164	.2401	.1975	.1785	.1296	.0915	.0677	.0625
	1	.0388	.1715	.2916	.3685	.4096	.4219	.4116	.3951	.3845	.3456	.2995	.2600	.2500
	2	.0006	.0135	.0486	.0975	.1536	.2109	.2646	.2963	.3105	.3456	.3675	.3747	.3750
	3	.0000	.0005	.0036	.0115	.0256	.0469	.0756	.0988	.1115	.1536	.2005	.2400	.2500
	4	.0000	.0000	.0001	.0005	.0016	.0039	.0081	.0123	.0150	.0256	.0410	.0576	.0625
5	0	.9510	.7738	.5905	.4437	.3277	.2373	.1681	.1317	.1160	.0778	.0503	.0345	.0312
	1	.0480	.2036	.3280	.3915	.4096	.3955	.3602	.3292	.3124	.2592	.2059	.1657	.1562
	2	.0010	.0214	.0729	.1382	.2048	.2637	.3087	.3292	.3364	.3456	.3369	.3185	.3125
	3	.0000	.0011	.0081	.0244	.0512	.0879	.1323	.1646	.1811	.2304	.2757	.3060	.3125
	4	.0000	.0000	.0004	.0022	.0064	.0146	.0284	.0412	.0488	.0768	.1128	.1470	.1562
	5	.0000	.0000	.0000	.0001	.0003	.0010	.0024	.0041	.0053	.0102	.0185	.0283	.0312
6	0	.9415	.7351	.5314	.3771	.2621	.1780	.1176	.0878	.0754	.0467	.0277	.0176	.0156
	1	.0571	.2321	.3543	.3993	.3932	.3560	.3025	.2634	.2437	.1866	.1359	.1014	.0938
	2	.0014	.0305	.0984	.1762	.2458	.2966	.3241	.3292	.3280	.3110	.2780	.2437	.2344
	3	.0000	.0021	.0146	.0415	.0819	.1318	.1852	.2195	.2355	.2765	.3032	.3121	.3125
	4	.0000	.0001	.0012	.0055	.0154	.0330	.0595	.0823	.0951	.1382	.1861	.2249	.2344
	5	.0000	.0000	.0001	.0004	.0015	.0044	.0102	.0165	.0205	.0369	.0609	.0864	.0938
	6	.0000	.0000	.0000	.0000	.0001	.0002	.0007	.0014	.0018	.0041	.0083	.0139	.0156

7	0	.9321	.6983	.4783	.3206	.2097	.1335	.0824	.0585	.0490	.0280	.0152	.0090	.0078
	1	.0659	.2573	.3720	.3960	.3670	.3115	.2471	.2048	.1848	.1306	.0872	.0603	.0547
	2	.0020	.0406	.1240	.2097	.2753	.3115	.3177	.3073	.2985	.2613	.2140	.1740	.1641
	3	.0000	.0036	.0230	.0617	.1147	.1730	.2269	.2561	.2679	.2903	.2918	.2786	.2734
	4	.0000	.0002	.0026	.0109	.0287	.0577	.0972	.1280	.1442	.1935	.2388	.2676	.2734
	5	.0000	.0000	.0002	.0012	.0043	.0115	.0250	.0384	.0466	.0774	.1172	.1543	.1641
	6	.0000	.0000	.0000	.0001	.0004	.0013	.0036	.0064	.0084	.0172	.0320	.0494	.0547
	7	.0000	.0000	.0000	.0000	.0000	.0001	.0002	.0005	.0006	.0016	.0037	.0068	.0078
8	0	.9227	.6634	.4305	.2725	.1678	.1001	.0576	.0390	.0319	.0168	.0084	.0046	.0039
	1	.0746	.2793	.3826	.3847	.3355	.2670	.1977	.1561	.1373	.0896	.0548	.0352	.0312
	2	.0026	.0515	.1488	.2376	.2936	.3115	.2965	.2731	.2587	.2090	.1569	.1183	.1094
	3	.0001	.0054	.0331	.0839	.1468	.2076	.2541	.2731	.2786	.2787	.2568	.2273	.2188
	4	.0000	.0004	.0046	.0185	.0459	.0865	.1361	.1707	.1875	.2322	.2627	.2730	.2734
	5	.0000	.0000	.0004	.0026	.0092	.0231	.0467	.0683	.0808	.1239	.1719	.2098	.2188
	6	.0000	.0000	.0000	.0002	.0011	.0038	.0100	.0171	.0217	.0413	.0703	.1008	.1094
	7	.0000	.0000	.0000	.0000	.0001	.0004	.0012	.0024	.0033	.0079	.0164	.0277	.0312
	8	.0000	.0000	.0000	.0000	.0000	.0000	.0001	.0002	.0002	.0007	.0017	.0033	.0039
9	0	.9135	.6302	.3874	.2316	.1342	.0751	.0404	.0260	.0207	.0101	.0046	.0023	.0020
	1	.0830	.2985	.3874	.3679	.3020	.2253	.1556	.1171	.1004	.0605	.0339	.0202	.0176
	2	.0034	.0629	.1722	.2597	.3020	.3003	.2668	.2341	.2162	.1612	.1110	.0776	.0703
	3	.0001	.0077	.0446	.1069	.1762	.2336	.2668	.2731	.2716	.2508	.2119	.1739	.1641
	4	.0000	.0006	.0074	.0283	.0661	.1168	.1715	.2048	.2194	.2508	.2600	.2506	.2461
	5	.0000	.0000	.0008	.0050	.0165	.0389	.0735	.1024	.1181	.1672	.2128	.2408	.2461
	6	.0000	.0000	.0001	.0006	.0028	.0087	.0210	.0341	.0424	.0743	.1160	.1542	.1641
	7	.0000	.0000	.0000	.0000	.0003	.0012	.0039	.0073	.0098	.0212	.0407	.0635	.0703
	8	.0000	.0000	.0000	.0000	.0000	.0001	.0004	.0009	.0013	.0035	.0083	.0153	.0176
	9	.0000	.0000	.0000	.0000	.0000	.0000	.0000	.0001	.0001	.0003	.0008	.0016	.0020
10	0	.9044	.5987	.3487	.1969	.1074	.0563	.0282	.0173	.0135	.0060	.0025	.0012	.0010
	1	.0914	.3151	.3874	.3474	.2684	.1877	.1211	.0867	.0725	.0403	.0207	.0114	.0098
	2	.0042	.0746	.1937	.2759	.3020	.2816	.2335	.1951	.1757	.1209	.0763	.0495	.0439
	3	.0001	.0105	.0574	.1298	.2013	.2503	.2668	.2601	.2522	.2150	.1665	.1267	.1172
	4	.0000	.0010	.0112	.0401	.0881	.1460	.2001	.2276	.2377	.2508	.2384	.2130	.2051
	5	.0000	.0001	.0015	.0085	.0264	.0584	.1029	.1366	.1536	.2007	.2340	.2456	.2461
	6	.0000	.0000	.0001	.0012	.0055	.0162	.0368	.0569	.0689	.1115	.1596	.1966	.2051
	7	.0000	.0000	.0000	.0001	.0008	.0031	.0090	.0163	.0212	.0425	.0746	.1080	.1172
	8	.0000	.0000	.0000	.0000	.0001	.0004	.0014	.0030	.0043	.0106	.0229	.0389	.0439
	9	.0000	.0000	.0000	.0000	.0000	.0000	.0001	.0003	.0005	.0016	.0042	.0083	.0098
	10	.0000	.0000	.0000	.0000	.0000	.0000	.0000	.0000	.0000	.0001	.0003	.0008	.0010

Source. Reprinted from Table II, *Modern Probability Theory and Its Applications* by E. Parzen. Copyright © 1960 by John Wiley & Sons, Inc., New York. Reprinted by permission of John Wiley & Sons, Inc.

Table A.3 Cumulative Normal Distribution

$$F(x)_{x=X} = \frac{1}{\sqrt{2\pi}} \int_{-\infty}^{X} e^{-u^2/2}\, du$$

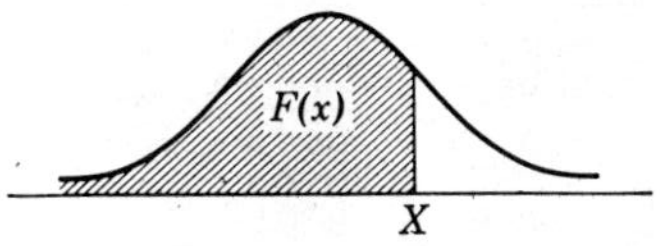

X	0.00	0.01	0.02	0.03	0.04	0.05	0.06	0.07	0.08	0.09
0.0	0.50000	0.50399	0.50798	0.51197	0.51595	0.51994	0.52392	0.52790	0.53188	0.53586
0.1	0.53983	0.54380	0.54776	0.55172	0.55567	0.55962	0.56356	0.56749	0.57142	0.57535
0.2	0.57926	0.58317	0.58706	0.59095	0.59483	0.59871	0.60257	0.60642	0.61026	0.61409
0.3	0.61791	0.62172	0.62552	0.62930	0.63307	0.63683	0.64058	0.64431	0.64803	0.65173
0.4	0.65542	0.65910	0.66276	0.66640	0.67003	0.67364	0.67724	0.68082	0.68439	0.68793
0.5	0.69146	0.69497	0.69847	0.70194	0.70540	0.70884	0.71226	0.71566	0.71904	0.72240
0.6	0.72575	0.72907	0.73237	0.73565	0.73891	0.74215	0.74537	0.74857	0.75175	0.75490
0.7	0.75804	0.76115	0.76424	0.76730	0.77035	0.77337	0.77637	0.77935	0.78230	0.78524
0.8	0.78814	0.79103	0.79389	0.79673	0.79955	0.80234	0.80511	0.80785	0.81057	0.81327
0.9	0.81594	0.81859	0.82121	0.82381	0.82639	0.82894	0.83147	0.83398	0.83646	0.83891
1.0	0.84134	0.84375	0.84614	0.84850	0.85083	0.85314	0.85543	0.85769	0.85993	0.86214
1.1	0.86433	0.86650	0.86864	0.87076	0.87286	0.87493	0.87698	0.87900	0.88100	0.88298
1.2	0.88493	0.88686	0.88877	0.89065	0.89251	0.89435	0.89617	0.89796	0.89973	0.90147
1.3	0.90320	0.90490	0.90658	0.90824	0.90988	0.91149	0.91309	0.91466	0.91621	0.91774
1.4	0.91924	0.92073	0.92220	0.92364	0.92507	0.92647	0.92786	0.92922	0.93056	0.93189
1.5	0.93319	0.93448	0.93574	0.93699	0.93822	0.93943	0.94062	0.94179	0.94295	0.94408
1.6	0.94520	0.94630	0.94738	0.94845	0.94950	0.95053	0.95154	0.95254	0.95352	0.95449
1.7	0.95543	0.95637	0.95728	0.95818	0.95907	0.95994	0.96080	0.96164	0.96246	0.96327
1.8	0.96407	0.96485	0.96562	0.96638	0.96712	0.96784	0.96856	0.96926	0.96995	0.97062
1.9	0.97128	0.97193	0.97257	0.97320	0.97381	0.97441	0.97500	0.97558	0.97615	0.97670

2.0	0.97725	0.97778	0.97831	0.97882	0.97932	0.97982	0.98030	0.98077	0.98124	0.98169
2.1	0.98214	0.98257	0.98300	0.98341	0.98382	0.98422	0.98461	0.98500	0.98537	0.98574
2.2	0.98610	0.98645	0.98679	0.98713	0.98745	0.98778	0.98809	0.98840	0.98870	0.98899
2.3	0.98928	0.98956	0.98983	0.99010	0.99036	0.99061	0.99086	0.99111	0.99134	0.99158
2.4	0.99180	0.99202	0.99224	0.99245	0.99266	0.99286	0.99305	0.99324	0.99343	0.99361
2.5	0.99379	0.99396	0.99413	0.99430	0.99446	0.99461	0.99477	0.99492	0.99506	0.99520
2.6	0.99534	0.99547	0.99560	0.99573	0.99585	0.99598	0.99609	0.99621	0.99632	0.99643
2.7	0.99653	0.99664	0.99674	0.99683	0.99693	0.99702	0.99711	0.99720	0.99728	0.99736
2.8	0.99744	0.99752	0.99760	0.99767	0.99774	0.99781	0.99788	0.99795	0.99801	0.99807
2.9	0.99813	0.99819	0.99825	0.99831	0.99836	0.99841	0.99846	0.99851	0.99856	0.99861
3.0	0.99865	0.99869	0.99874	0.99878	0.99882	0.99886	0.99889	0.99893	0.99897	0.99900
3.1	0.99903	0.99906	0.99910	0.99913	0.99916	0.99918	0.99921	0.99924	0.99926	0.99929
3.2	0.99931	0.99934	0.99936	0.99938	0.99940	0.99942	0.99944	0.99946	0.99948	0.99950
3.3	0.99952	0.99953	0.99957	0.99957	0.99958	0.99960	0.99961	0.99962	0.99964	0.99965
3.4	0.99966	0.99968	0.99969	0.99970	0.99971	0.99972	0.99973	0.99974	0.99975	0.99976
3.5	0.99977	0.99978	0.99978	0.99979	0.99980	0.99981	0.99981	0.99982	0.99983	0.99983
3.6	0.99984	0.99985	0.99985	0.99986	0.99986	0.99987	0.99987	0.99988	0.99988	0.99989
3.7	0.99989	0.99990	0.99990	0.99990	0.99991	0.99991	0.99992	0.99992	0.99992	0.99992
3.8	0.99993	0.99993	0.99993	0.99994	0.99994	0.99994	0.99994	0.99995	0.99995	0.99995
3.9	0.99995	0.99995	0.99996	0.99996	0.99996	0.99996	0.99996	0.99996	0.99997	0.99997

Source. Reprinted from Table C, *Statistics and Experimental Design in Engineering and the Physical Sciences* by N. L. Johnson and F. C. Leone. Copyright © 1964 by John Wiley & Sons, Inc., New York. Also reprinted from *Biometrika Tables for Statisticians,* Vol. 1, by E. S. Pearson and H. O. Hartley, pp. 104–108. Copyright © 1958 by Biometrika Trustees. Reprinted by permission of John Wiley & Sons, Inc. and the Biometrika Trustees.

Table A.4 Poisson Probabilities

A table of $e^{-\lambda}\lambda^x/x!$ for $\lambda = 0.1(0.1)2(0.2)4(1)10$

λ \ x	0	1	2	3	4	5	6	7	8	9	10	11
.1	.9048	.0905	.0045	.0002	.0000							
.2	.8187	.1637	.0164	.0011	.0001	.0000						
.3	.7408	.2222	.0333	.0033	.0002	.0000						
.4	.6703	.2681	.0536	.0072	.0007	.0001	.0000					
.5	.6065	.3033	.0758	.0126	.0016	.0002	.0000					
.6	.5488	.3293	.0988	.0198	.0030	.0004	.0000					
.7	.4966	.3476	.1217	.0284	.0050	.0007	.0001	.0000				
.8	.4493	.3595	.1438	.0383	.0077	.0012	.0002	.0000				
.9	.4066	.3659	.1647	.0494	.0111	.0020	.0003	.0000				
1.0	.3679	.3679	.1839	.0613	.0153	.0031	.0005	.0001	.0000			
1.1	.3329	.3662	.2014	.0738	.0203	.0045	.0008	.0001	.0000			
1.2	.3012	.3614	.2169	.0867	.0260	.0062	.0012	.0002	.0000			
1.3	.2725	.3543	.2303	.0998	.0324	.0084	.0018	.0003	.0001	.0000		
1.4	.2466	.3452	.2417	.1128	.0395	.0111	.0026	.0005	.0001	.0000		
1.5	.2231	.3347	.2510	.1255	.0471	.0141	.0035	.0008	.0001	.0000		
1.6	.2019	.3230	.2584	.1378	.0551	.0176	.0047	.0011	.0002	.0000		
1.7	.1827	.3106	.2640	.1496	.0636	.0216	.0061	.0015	.0003	.0001	.0000	
1.8	.1653	.2975	.2678	.1607	.0723	.0260	.0078	.0020	.0005	.0001	.0000	
1.9	.1496	.2842	.2700	.1710	.0812	.0309	.0098	.0027	.0006	.0001	.0000	
2.0	.1353	.2707	.2707	.1804	.0902	.0361	.0120	.0034	.0009	.0002	.0000	
2.2	.1108	.2438	.2681	.1966	.1082	.0476	.0174	.0055	.0015	.0004	.0001	.0000
2.4	.0907	.2177	.2613	.2090	.1254	.0602	.0241	.0083	.0025	.0007	.0002	.0000
2.6	.0743	.1931	.2510	.2176	.1414	.0735	.0319	.0118	.0038	.0011	.0003	.0001
2.8	.0608	.1703	.2384	.2225	.1557	.0872	.0407	.0163	.0057	.0018	.0005	.0001
3.0	.0498	.1494	.2240	.2240	.1680	.1008	.0504	.0216	.0081	.0027	.0008	.0002
3.2	.0408	.1304	.2087	.2226	.1781	.1140	.0608	.0278	.0111	.0040	.0013	.0004
3.4	.0334	.1135	.1929	.2186	.1858	.1264	.0716	.0348	.0148	.0056	.0019	.0006
3.6	.0273	.0984	.1771	.2125	.1912	.1377	.0826	.0425	.0191	.0076	.0028	.0009
3.8	.0224	.0850	.1615	.2046	.1944	.1477	.0936	.0508	.0241	.0102	.0039	.0013
4.0	.0183	.0733	.1465	.1954	.1954	.1563	.1042	.0595	.0298	.0132	.0053	.0019
5.0	.0067	.0337	.0842	.1404	.1755	.1755	.1462	.1044	.0653	.0363	.0181	.0082
6.0	.0025	.0149	.0446	.0892	.1339	.1606	.1606	.1377	.1033	.0688	.0413	.0225
7.0	.0009	.0064	.0223	.0521	.0912	.1277	.1490	.1490	.1304	.1014	.0710	.0452
8.0	.0003	.0027	.0107	.0286	.0573	.0916	.1221	.1396	.1396	.1241	.0993	.0722
9.0	.0001	.0011	.0050	.0150	.0337	.0607	.0911	.1171	.1318	.1318	.1186	.0970
10.0	.0000	.0005	.0023	.0076	.0189	.0378	.0631	.0901	.1126	.1251	.1251	.1137

Table A.4 *(Continued)*

λ \ x	12	13	14	15	16	17	18	19	20	21	22	23	24
2.6	.0000												
2.8	.0000												
3.0	.0001												
3.2	.0001												
3.4	.0002												
3.6	.0003												
3.8	.0004												
4.0	.0006												
5.0	.0034	.0013	.0005	.0002									
6.0	.0113	.0052	.0022	.0009	.0003	.0001							
7.0	.0264	.0142	.0071	.0033	.0014	.0006	.0002	.0001					
8.0	.0481	.0296	.0169	.0090	.0045	.0021	.0009	.0004	.0002	.0001			
9.0	.0728	.0504	.0324	.0194	.0109	.0058	.0029	.0014	.0006	.0003	.0001		
10.0	.0948	.0729	.0521	.0347	.0217	0128	.0071	.0037	.0019	.0009	.0004	.0002	.0001

Source. Reprinted from Table III, *Modern Probability Theory and Its Applications* by E. Parzen. Copyright © 1964 by John Wiley & Sons, Inc., New York. Reprinted by permission of John Wiley & Sons, Inc.

Table A.5 "Student's" t Distribution

Values of t_0 such that

$$P = \int_{-\infty}^{t_0} \frac{1}{\sqrt{\nu\pi}} \frac{\Gamma\left(\frac{\nu+1}{2}\right)}{\Gamma\left(\frac{\nu}{2}\right)} \left(1 + \frac{t^2}{\nu}\right)^{-\frac{\nu+1}{2}} dt$$

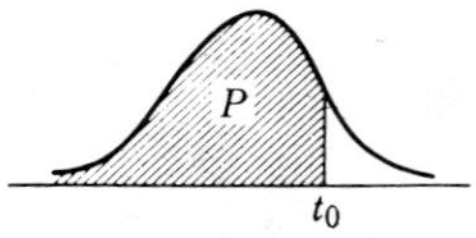

ν \ P	0.750	0.900	0.950	0.975	0.990	0.995	0.999	0.9995
1	1.000	3.078	6.314	12.706	31.821	63.657	318.31	636.62
2	0.816	1.886	2.920	4.303	6.965	9.925	22.326	31.598
3	0.765	1.638	2.353	3.182	4.541	5.841	10.213	12.924
4	0.741	1.533	2.132	2.776	3.747	4.604	7.173	8.610
5	0.727	1.476	2.015	2.571	3.365	4.032	5.893	6.869
6	0.718	1.440	1.943	2.447	3.143	3.707	5.208	5.959
7	0.711	1.415	1.895	2.365	2.998	3.499	4.785	5.408
8	0.706	1.397	1.860	2.306	2.896	3.355	4.501	5.041
9	0.703	1.383	1.833	2.262	2.821	3.250	4.297	4.781
10	0.700	1.372	1.812	2.228	2.764	3.169	4.144	4.587
11	0.697	1.363	1.796	2.201	2.718	3.106	4.025	4.437
12	0.695	1.356	1.782	2.179	2.681	3.055	3.930	4.318
13	0.694	1.350	1.771	2.160	2.650	3.012	3.852	4.221
14	0.692	1.345	1.761	2.145	2.624	2.977	3.787	4.140
15	0.691	1.341	1.753	2.131	2.602	2.947	3.733	4.073
16	0.690	1.337	1.746	2.120	2.583	2.921	3.686	4.015
17	0.689	1.333	1.740	2.110	2.567	2.898	3.646	3.965
18	0.688	1.330	1.734	2.101	2.552	2.878	3.610	3.922
19	0.688	1.328	1.729	2.093	2.539	2.861	3.579	3.883
20	0.687	1.325	1.725	2.086	2.528	2.845	3.552	3.850
21	0.686	1.323	1.721	2.080	2.518	2.831	3.527	3.819
22	0.686	1.321	1.717	2.074	2.508	2.819	3.505	3.792
23	0.685	1.319	1.714	2.069	2.500	2.807	3.485	3.767
24	0.685	1.318	1.711	2.064	2.492	2.797	3.467	3.745
25	0.684	1.316	1.708	2.060	2.485	2.787	3.450	3.725
26	0.684	1.315	1.706	2.056	2.479	2.779	3.435	3.707
27	0.684	1.314	1.703	2.052	2.473	2.771	3.421	3.690
28	0.683	1.313	1.701	2.048	2.467	2.763	3.408	3.674
29	0.683	1.311	1.699	2.045	2.462	2.756	3.396	3.659
30	0.683	1.310	1.697	2.042	2.457	2.750	3.385	3.646
40	0.681	1.303	1.684	2.021	2.423	2.704	3.307	3.551
60	0.679	1.296	1.671	2.000	2.390	2.660	3.232	3.460
120	0.677	1.289	1.658	1.980	2.358	2.617	3.160	3.373
∞	0.674	1.282	1.645	1.960	2.326	2.576	3.090	3.291

Source. Reprinted from Table F, *Statistics and Experimental Design in Engineering and the Physical Sciences* by N. L. Johnson and F. C. Leone. Copyright © 1964 by John Wiley & Sons, Inc. From Pearson and Hartley, *Biometrika Tables for Statisticians,* Vol. 1 (1958), p. 138, and from Table III, Fisher and Yates, *Statistical Tables for Biological, Agricultural and Medical Research,* published by Longman Group Ltd., London. (Previously published by Oliver & Boyd, Edinburgh) and by permission of authors and publishers.

Table A.6 Values of the Correlation Coefficient for Different Levels of Significance

n	.1	.05	.02	.01	.001	n	.1	.05	.02	.01	.001
1	.98769	.99692	.999507	.999877	.9999988	16	.4000	.4683	.5425	.5897	.7084
2	.90000	.95000	.98000	.990000	.99900	17	.3887	.4555	.5285	.5751	.6932
3	.8054	.8783	.93433	.95873	.99116	18	.3783	.4438	.5155	.5614	.6787
4	.7293	.8114	.8822	.91720	.97406	19	.3687	.4329	.5034	.5487	.6652
5	.6694	.7545	.8329	.8745	.95074	20	.3598	.4227	.4921	.5368	.6524
6	.6215	.7067	.7887	.8343	.92493	25	.3233	.3809	.4451	.4869	.5974
7	.5822	.6664	.7498	.7977	.8982	30	.2960	.3494	.4093	.4487	.5541
8	.5494	.6319	.7155	.7646	.8721	35	.2746	.3246	.3810	.4182	.5189
9	.5214	.6021	.6851	.7348	.8471	40	.2573	.3044	.3578	.3932	.4896
10	.4973	.5760	.6581	.7079	.8233	45	.2428	.2875	.3384	.3721	.4648
11	.4762	.5529	.6339	.6835	.8010	50	.2306	.2732	.3218	.3541	.4433
12	.4575	.5324	.6120	.6614	.7800	60	.2108	.2500	.2948	.3248	.4078
13	.4409	.5139	.5923	.6411	.7603	70	.1954	.2319	.2737	.3017	.3799
14	.4259	.4973	.5742	.6226	.7420	80	.1829	.2172	.2565	.2830	.3568
15	.4124	.4821	.5577	.6055	.7246	90	.1726	.2050	.2422	.2673	.3375
						100	.1638	.1946	.2301	.2540	.3211

Source. Reproduced by permission of the publishers, Longman Group Limited, London, from Fisher & Yates, *Statistical Tables for Biological, Agricultural, and Medical Research,* 6th ed., 1974.

Table A.7 The F Distribution

Values of $F_{\nu_1,\nu_2,P}$ such that $P = \dfrac{1}{B(\nu_1/2,\ \nu_2/2)} \displaystyle\int_0^{\nu_1 F_{\nu_1,\nu_2,P}/\nu_2} g^{\nu_1/2-1}(1+g)^{-(\nu_1+\nu_2)/2}\,dg$

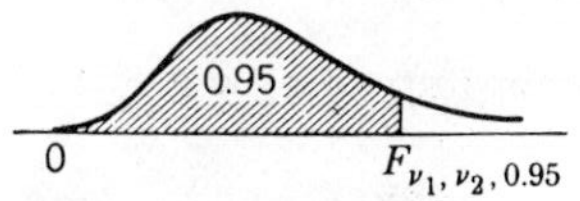

$P = 0.95$

ν_2 \ ν_1	1	2	3	4	5	6	7	8	9	10	12	15	20	24	30	40	60	120	∞
1	161.4	199.5	215.7	224.6	230.2	234.0	236.8	238.9	240.5	241.9	243.9	245.9	248.0	249.1	250.1	251.1	252.2	253.3	254.3
2	18.51	19.00	19.16	19.25	19.30	19.33	19.35	19.37	19.38	19.40	19.41	19.43	19.45	19.45	19.46	19.47	19.48	19.49	19.50
3	10.13	9.55	9.28	9.12	9.01	8.94	8.89	8.85	8.81	8.79	8.74	8.70	8.66	8.64	8.62	8.59	8.57	8.55	8.53
4	7.71	6.94	6.59	6.39	6.26	6.16	6.09	6.04	6.00	5.96	5.91	5.86	5.80	5.77	5.75	5.72	5.69	5.66	5.63
5	6.61	5.79	5.41	5.19	5.05	4.95	4.88	4.82	4.77	4.74	4.68	4.62	4.56	4.53	4.50	4.46	4.43	4.40	4.36
6	5.99	5.14	4.76	4.53	4.39	4.28	4.21	4.15	4.10	4.06	4.00	3.94	3.87	3.84	3.81	3.77	3.74	3.70	3.67
7	5.59	4.74	4.35	4.12	3.97	3.87	3.79	3.73	3.68	3.64	3.57	3.51	3.44	3.41	3.38	3.34	3.30	3.27	3.23
8	5.32	4.46	4.07	3.84	3.69	3.58	3.50	3.44	3.39	3.35	3.28	3.22	3.15	3.12	3.08	3.04	3.01	2.97	2.93
9	5.12	4.26	3.86	3.63	3.48	3.37	3.29	3.23	3.18	3.14	3.07	3.01	2.94	2.90	2.86	2.83	2.79	2.75	2.71
10	4.96	4.10	3.71	3.48	3.33	3.22	3.14	3.07	3.02	2.98	2.91	2.85	2.77	2.74	2.70	2.66	2.62	2.58	2.54
11	4.84	3.98	3.59	3.36	3.20	3.09	3.01	2.95	2.90	2.85	2.79	2.72	2.65	2.61	2.57	2.53	2.49	2.45	2.40
12	4.75	3.89	3.49	3.26	3.11	3.00	2.91	2.85	2.80	2.75	2.69	2.62	2.54	2.51	2.47	2.43	2.38	2.34	2.30
13	4.67	3.81	3.41	3.18	3.03	2.92	2.83	2.77	2.71	2.67	2.60	2.53	2.46	2.42	2.38	2.34	2.30	2.25	2.21
14	4.60	3.74	3.34	3.11	2.96	2.85	2.76	2.70	2.65	2.60	2.53	2.46	2.39	2.35	2.31	2.27	2.22	2.18	2.13
15	4.54	3.68	3.29	3.06	2.90	2.79	2.71	2.64	2.59	2.54	2.48	2.40	2.33	2.29	2.25	2.20	2.16	2.11	2.07
16	4.49	3.63	3.24	3.01	2.85	2.74	2.66	2.59	2.54	2.49	2.42	2.35	2.28	2.24	2.19	2.15	2.11	2.06	2.01
17	4.45	3.59	3.20	2.96	2.81	2.70	2.61	2.55	2.49	2.45	2.38	2.31	2.23	2.19	2.15	2.10	2.06	2.01	1.96
18	4.41	3.55	3.16	2.93	2.77	2.66	2.58	2.51	2.46	2.41	2.34	2.27	2.19	2.15	2.11	2.06	2.02	1.97	1.92
19	4.38	3.52	3.13	2.90	2.74	2.63	2.54	2.48	2.42	2.38	2.31	2.23	2.16	2.11	2.07	2.03	1.98	1.93	1.88
20	4.35	3.49	3.10	2.87	2.71	2.60	2.51	2.45	2.39	2.35	2.28	2.20	2.12	2.08	2.04	1.99	1.95	1.90	1.84
21	4.32	3.47	3.07	2.84	2.68	2.57	2.49	2.42	2.37	2.32	2.25	2.18	2.10	2.05	2.01	1.96	1.92	1.87	1.81
22	4.30	3.44	3.05	2.82	2.66	2.55	2.46	2.40	2.34	2.30	2.23	2.15	2.07	2.03	1.98	1.94	1.89	1.84	1.78
23	4.28	3.42	3.03	2.80	2.64	2.53	2.44	2.37	2.32	2.27	2.20	2.13	2.05	2.01	1.96	1.91	1.86	1.81	1.76
24	4.26	3.40	3.01	2.78	2.62	2.51	2.42	2.36	2.30	2.25	2.18	2.11	2.03	1.98	1.94	1.89	1.84	1.79	1.73
25	4.24	3.39	2.99	2.76	2.60	2.49	2.40	2.34	2.28	2.24	2.16	2.09	2.01	1.96	1.92	1.87	1.82	1.77	1.71
26	4.23	3.37	2.98	2.74	2.59	2.47	2.39	2.32	2.27	2.22	2.15	2.07	1.99	1.95	1.90	1.85	1.80	1.75	1.69
27	4.21	3.35	2.96	2.73	2.57	2.46	2.37	2.31	2.25	2.20	2.13	2.06	1.97	1.93	1.88	1.84	1.79	1.73	1.67
28	4.20	3.34	2.95	2.71	2.56	2.45	2.36	2.29	2.24	2.19	2.12	2.04	1.96	1.91	1.87	1.82	1.77	1.71	1.65
29	4.18	3.33	2.93	2.70	2.55	2.43	2.35	2.28	2.22	2.18	2.10	2.03	1.94	1.90	1.85	1.81	1.75	1.70	1.64
30	4.17	3.32	2.92	2.69	2.53	2.42	2.33	2.27	2.21	2.16	2.09	2.01	1.93	1.89	1.84	1.79	1.74	1.68	1.62
40	4.08	3.23	2.84	2.61	2.45	2.34	2.25	2.18	2.12	2.08	2.00	1.92	1.84	1.79	1.74	1.69	1.64	1.58	1.51
60	4.00	3.15	2.76	2.53	2.37	2.25	2.17	2.10	2.04	1.99	1.92	1.84	1.75	1.70	1.65	1.59	1.53	1.47	1.39
120	3.92	3.07	2.68	2.45	2.29	2.17	2.09	2.02	1.96	1.91	1.83	1.75	1.66	1.61	1.55	1.50	1.43	1.35	1.25
∞	3.84	3.00	2.60	2.37	2.21	2.10	2.01	1.94	1.88	1.83	1.75	1.67	1.57	1.52	1.46	1.39	1.32	1.22	1.00

Source. Reprinted from Table G, *Statistics and Experimental Design in Engineering and the Physical Sciences* by N. L. Johnson and F. C. Leone. Copyright © 1964 by John Wiley & Sons, Inc., New York. Also reprinted from *Biometrika Tables for Statisticians,* Vol. I, by E. S. Pearson and H. O. Hartley, pp. 159–163. Copyright © 1958. Reprinted by permission of John Wiley & Sons, Inc. and the Biometrika Trustees.

Table A.7 (Continued)

$P = 0.975$

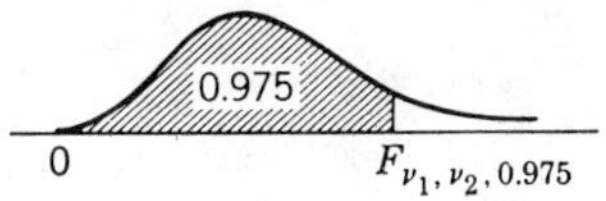

ν_2 \ ν_1	1	2	3	4	5	6	7	8	9	10	12	15	20	24	30	40	60	120	∞
1	647.8	799.5	864.2	899.6	921.8	937.1	948.2	956.7	963.3	968.6	976.7	984.9	993.1	997.2	1001	1006	1010	1014	1018
2	38.51	39.00	39.17	39.25	39.30	39.33	39.36	39.37	39.39	39.40	39.41	39.43	39.45	39.46	39.46	39.47	39.48	39.49	39.50
3	17.44	16.04	15.44	15.10	14.88	14.73	14.62	14.54	14.47	14.42	14.34	14.25	14.17	14.12	14.08	14.04	13.99	13.95	13.90
4	12.22	10.65	9.98	9.60	9.36	9.20	9.07	8.98	8.90	8.84	8.75	8.66	8.56	8.51	8.46	8.41	8.36	8.31	8.26
5	10.01	8.43	7.76	7.39	7.15	6.98	6.85	6.76	6.68	6.62	6.52	6.43	6.33	6.28	6.23	6.18	6.12	6.07	6.02
6	8.81	7.26	6.60	6.23	5.99	5.82	5.70	5.60	5.52	5.46	5.37	5.27	5.17	5.12	5.07	5.01	4.96	4.90	4.85
7	8.07	6.54	5.89	5.52	5.29	5.12	4.99	4.90	4.82	4.76	4.67	4.57	4.47	4.42	4.36	4.31	4.25	4.20	4.14
8	7.57	6.06	5.42	5.05	4.82	4.65	4.53	4.43	4.36	4.30	4.20	4.10	4.00	3.95	3.89	3.84	3.78	3.73	3.67
9	7.21	5.71	5.08	4.72	4.48	4.32	4.20	4.10	4.03	3.96	3.87	3.77	3.67	3.61	3.56	3.51	3.45	3.39	3.33
10	6.94	5.46	4.83	4.47	4.24	4.07	3.95	3.85	3.78	3.72	3.62	3.52	3.42	3.37	3.31	3.26	3.20	3.14	3.08
11	6.72	5.26	4.63	4.28	4.04	3.88	3.76	3.66	3.59	3.53	3.43	3.33	3.23	3.17	3.12	3.06	3.00	2.94	2.88
12	6.55	5.10	4.47	4.12	3.88	3.73	3.61	3.51	3.44	3.37	3.28	3.18	3.07	3.02	2.96	2.91	2.85	2.79	2.72
13	6.41	4.97	4.35	4.00	3.77	3.60	3.48	3.39	3.31	3.25	3.15	3.05	2.95	2.89	2.84	2.78	2.72	2.66	2.60
14	6.30	4.86	4.24	3.89	3.66	3.50	3.38	3.29	3.21	3.15	3.05	2.95	2.84	2.79	2.73	2.67	2.61	2.55	2.49
15	6.20	4.77	4.15	3.80	3.58	3.41	3.29	3.20	3.12	3.06	2.96	2.86	2.76	2.70	2.64	2.59	2.52	2.46	2.40
16	6.12	4.69	4.08	3.73	3.50	3.34	3.22	3.12	3.05	2.99	2.89	2.79	2.68	2.63	2.57	2.51	2.45	2.38	2.32
17	6.04	4.62	4.01	3.66	3.44	3.28	3.16	3.06	2.98	2.92	2.82	2.72	2.62	2.56	2.50	2.44	2.38	2.32	2.25
18	5.98	4.56	3.95	3.61	3.38	3.22	3.10	3.01	2.93	2.87	2.77	2.67	2.56	2.50	2.44	2.38	2.32	2.26	2.19
19	5.92	4.51	3.90	3.56	3.33	3.17	3.05	2.96	2.88	2.82	2.72	2.62	2.51	2.45	2.39	2.33	2.27	2.20	2.13
20	5.87	4.46	3.86	3.51	3.29	3.13	3.01	2.91	2.84	2.77	2.68	2.57	2.46	2.41	2.35	2.29	2.22	2.16	2.09
21	5.83	4.42	3.82	3.48	3.25	3.09	2.97	2.87	2.80	2.73	2.64	2.53	2.42	2.37	2.31	2.25	2.18	2.11	2.04
22	5.79	4.38	3.78	3.44	3.22	3.05	2.93	2.84	2.76	2.70	2.60	2.50	2.39	2.33	2.27	2.21	2.14	2.08	2.00
23	5.75	4.35	3.75	3.41	3.18	3.02	2.90	2.81	2.73	2.67	2.57	2.47	2.36	2.30	2.24	2.18	2.11	2.04	1.97
24	5.72	4.32	3.72	3.38	3.15	2.99	2.87	2.78	2.70	2.64	2.54	2.44	2.33	2.27	2.21	2.15	2.08	2.01	1.94
25	5.69	4.29	3.69	3.35	3.13	2.97	2.85	2.75	2.68	2.61	2.51	2.41	2.30	2.24	2.18	2.12	2.05	1.98	1.91
26	5.66	4.27	3.67	3.33	3.10	2.94	2.82	2.73	2.65	2.59	2.49	2.39	2.28	2.22	2.16	2.09	2.03	1.95	1.88
27	5.63	4.24	3.65	3.31	3.08	2.92	2.80	2.71	2.63	2.57	2.47	2.36	2.25	2.19	2.13	2.07	2.00	1.93	1.85
28	5.61	4.22	3.63	3.29	3.06	2.90	2.78	2.69	2.61	2.55	2.45	2.34	2.23	2.17	2.11	2.05	1.98	1.91	1.83
29	5.59	4.20	3.61	3.27	3.04	2.88	2.76	2.67	2.59	2.53	2.43	2.32	2.21	2.15	2.09	2.03	1.96	1.89	1.81
30	5.57	4.18	3.59	3.25	3.03	2.87	2.75	2.65	2.57	2.51	2.41	2.31	2.20	2.14	2.07	2.01	1.94	1.87	1.79
40	5.42	4.05	3.46	3.13	2.90	2.74	2.62	2.53	2.45	2.39	2.29	2.18	2.07	2.01	1.94	1.88	1.80	1.72	1.64
60	5.29	3.93	3.34	3.01	2.79	2.63	2.51	2.41	2.33	2.27	2.17	2.06	1.94	1.88	1.82	1.74	1.67	1.58	1.48
120	5.15	3.80	3.23	2.89	2.67	2.52	2.39	2.30	2.22	2.16	2.05	1.94	1.82	1.76	1.69	1.61	1.53	1.43	1.31
∞	5.02	3.69	3.12	2.79	2.57	2.41	2.29	2.19	2.11	2.05	1.94	1.83	1.71	1.64	1.57	1.48	1.39	1.27	1.00

Table A.7 (*Continued*)

$P = 0.99$

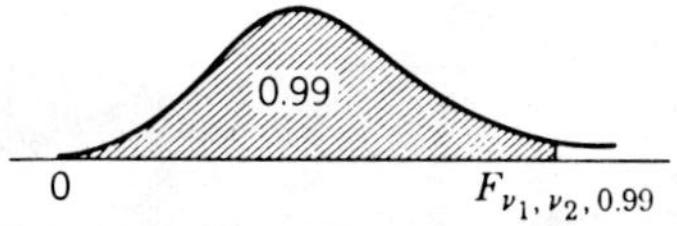

ν_2 \ ν_1	1	2	3	4	5	6	7	8	9	10	12	15	20	24	30	40	60	120	∞
1	4052	4999.5	5403	5625	5764	5859	5928	5982	6022	6056	6106	6157	6209	6235	6261	6287	6313	6339	6366
2	98.50	99.00	99.17	99.25	99.30	99.33	99.36	99.37	99.39	99.40	99.42	99.43	99.45	99.46	99.47	99.47	99.48	99.49	99.50
3	34.12	30.82	29.46	28.71	28.24	27.91	27.67	27.49	27.35	27.23	27.05	26.87	26.69	26.60	26.50	26.41	26.32	26.22	26.13
4	21.20	18.00	16.69	15.98	15.52	15.21	14.98	14.80	14.66	14.55	14.37	14.20	14.02	13.93	13.84	13.75	13.65	13.56	13.46
5	16.26	13.27	12.06	11.39	10.97	10.67	10.46	10.29	10.16	10.05	9.89	9.72	9.55	9.47	9.38	9.29	9.20	9.11	9.02
6	13.75	10.92	9.78	9.15	8.75	8.47	8.26	8.10	7.98	7.87	7.72	7.56	7.40	7.31	7.23	7.14	7.06	6.97	6.88
7	12.25	9.55	8.45	7.85	7.46	7.19	6.99	6.84	6.72	6.62	6.47	6.31	6.16	6.07	5.99	5.91	5.82	5.74	5.65
8	11.26	8.65	7.59	7.01	6.63	6.37	6.18	6.03	5.91	5.81	5.67	5.52	5.36	5.28	5.20	5.12	5.03	4.95	4.86
9	10.56	8.02	6.99	6.42	6.06	5.80	5.61	5.47	5.35	5.26	5.11	4.96	4.81	4.73	4.65	4.57	4.48	4.40	4.31
10	10.04	7.56	6.55	5.99	5.64	5.39	5.20	5.06	4.94	4.85	4.71	4.56	4.41	4.33	4.25	4.17	4.08	4.00	3.91
11	9.65	7.21	6.22	5.67	5.32	5.07	4.89	4.74	4.63	4.54	4.40	4.25	4.10	4.02	3.94	3.86	3.78	3.69	3.60
12	9.33	6.93	5.95	5.41	5.06	4.82	4.64	4.50	4.39	4.30	4.16	4.01	3.86	3.78	3.70	3.62	3.54	3.45	3.36
13	9.07	6.70	5.74	5.21	4.86	4.62	4.44	4.30	4.19	4.10	3.96	3.82	3.66	3.59	3.51	3.43	3.34	3.25	3.17
14	8.86	6.51	5.56	5.04	4.69	4.46	4.28	4.14	4.03	3.94	3.80	3.66	3.51	3.43	3.35	3.27	3.18	3.09	3.00
15	8.68	6.36	5.42	4.89	4.56	4.32	4.14	4.00	3.89	3.80	3.67	3.52	3.37	3.29	3.21	3.13	3.05	2.96	2.87
16	8.53	6.23	5.29	4.77	4.44	4.20	4.03	3.89	3.78	3.69	3.55	3.41	3.26	3.18	3.10	3.02	2.93	2.64	2.75
17	8.40	6.11	5.18	4.67	4.34	4.10	3.93	3.79	3.68	3.59	3.46	3.31	3.16	3.08	3.00	2.92	2.83	2.75	2.65
18	8.29	6.01	5.09	4.58	4.25	4.01	3.84	3.71	3.60	3.51	3.37	3.23	3.08	3.00	2.92	2.84	2.75	2.66	2.57
19	8.18	5.93	5.01	4.50	4.17	3.94	3.77	3.63	3.52	3.43	3.30	3.15	3.00	2.92	2.84	2.76	2.67	2.58	2.49
20	8.10	5.85	4.94	4.43	4.10	3.87	3.70	3.56	3.46	3.37	3.23	3.09	2.94	2.86	2.78	2.69	2.61	2.52	2.42
21	8.02	5.78	4.87	4.37	4.04	3.81	3.64	3.51	3.40	3.31	3.17	3.03	2.88	2.80	2.72	2.64	2.55	2.46	2.36
22	7.95	5.72	4.82	4.31	3.99	3.76	3.59	3.45	3.35	3.26	3.12	2.98	2.83	2.75	2.67	2.58	2.50	2.40	2.31
23	7.88	5.66	4.76	4.26	3.94	3.71	3.54	3.41	3.30	3.21	3.07	2.93	2.78	2.70	2.62	2.54	2.45	2.35	2.26
24	7.82	5.61	4.72	4.22	3.90	3.67	3.50	3.36	3.26	3.17	3.03	2.89	2.74	2.66	2.58	2.49	2.40	2.31	2.21
25	7.77	5.57	4.68	4.18	3.85	3.63	3.46	3.32	3.22	3.13	2.99	2.85	2.70	2.62	2.54	2.45	2.36	2.27	2.17
26	7.72	5.53	4.64	4.14	3.82	3.59	3.42	3.29	3.18	3.09	2.96	2.81	2.66	2.58	2.50	2.42	2.33	2.23	2.13
27	7.68	5.49	4.60	4.11	3.78	3.56	3.39	3.26	3.15	3.06	2.93	2.78	2.63	2.55	2.47	2.38	2.29	2.20	2.10
28	7.64	5.45	4.57	4.07	3.75	3.53	3.36	3.23	3.12	3.03	2.90	2.75	2.60	2.52	2.44	2.35	2.26	2.17	2.06
29	7.60	5.42	4.54	4.04	3.73	3.50	3.33	3.20	3.09	3.00	2.87	2.73	2.57	2.49	2.41	2.33	2.23	2.14	2.03
30	7.56	5.39	4.51	4.02	3.70	3.47	3.30	3.17	3.07	2.98	2.84	2.70	2.55	2.47	2.39	2.30	221	2.11	2.01
40	7.31	5.18	4.31	3.83	3.51	3.29	3.12	2.99	2.89	2.80	2.66	2.52	2.37	2.29	2.20	2.11	2.02	1.92	1.80
60	7.08	4.98	4.13	3.65	3.34	3.12	2.95	2.82	2.72	2.63	2.50	2.35	2.20	2.12	2.03	1.94	1.84	1.70	1.60
120	6.85	4.79	3.95	3.48	3.17	2.96	2.79	2.66	2.56	2.47	2.34	2.19	2.03	1.95	1.86	1.76	1.66	1.53	1.38
∞	6.63	4.61	3.78	3.32	3.02	2.80	2.64	2.51	2.41	2.32	2.18	2.04	1.88	1.79	1.70	1.59	1.47	1.32	1.00

Table A.7 (*Continued*)

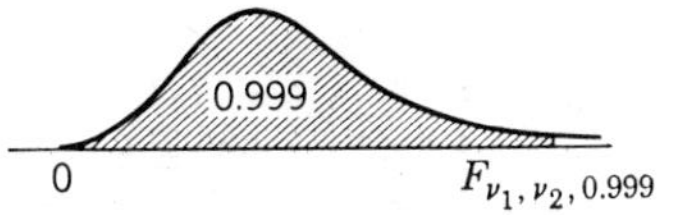

$P = 0.999$

ν_2 \ ν_1	1	2	3	4	5	6	7	8	9	10	12	15	20	24	30	40	60	120	∞
1	4053[a]	5000[a]	5404[a]	5625[a]	5764[a]	5859[a]	5929[a]	5981[a]	6023[a]	6056[a]	6107[a]	6158[a]	6209[a]	6235[a]	6261[a]	6287[a]	6313[a]	6340[a]	6366[a]
2	998.5	999.0	999.2	999.2	999.3	999.3	999.4	999.4	999.4	999.4	999.4	999.4	999.4	999.5	999.5	999.5	999.5	999.5	999.5
3	167.0	148.5	141.1	137.1	134.6	132.8	131.6	130.6	129.9	129.2	128.3	127.4	126.4	125.9	125.4	125.0	124.5	124.0	123.5
4	74.14	61.25	56.18	53.44	51.71	50.53	49.66	49.00	48.47	48.05	47.41	46.76	46.10	45.77	45.43	45.09	44.75	44.40	44.05
5	47.18	37.12	33.20	31.09	29.75	28.84	28.16	27.64	27.24	26.92	26.42	25.91	25.39	25.14	24.87	24.60	24.33	24.06	23.79
6	35.51	27.00	23.70	21.92	20.81	20.03	19.46	19.03	18.69	18.41	17.99	17.56	17.12	16.89	16.67	16.44	16.21	15.99	15.75
7	29.25	21.69	18.77	17.19	16.21	15.52	15.02	14.63	14.33	14.08	13.71	13.32	12.93	12.73	12.53	12.33	12.12	11.91	11.70
8	25.42	18.49	15.83	14.39	13.49	12.86	12.40	12.04	11.77	11.54	11.19	10.84	10.48	10.30	10.11	9.92	9.73	9.53	9.33
9	22.86	16.39	13.90	12.56	11.71	11.13	10.70	10.37	10.11	9.89	9.57	9.24	8.90	8.72	8.55	8.37	8.19	8.00	7.81
10	21.04	14.91	12.55	11.28	10.48	9.92	9.52	9.20	8.96	8.75	8.45	8.13	7.80	7.64	7.47	7.30	7.12	6.94	6.76
11	19.69	13.81	11.56	10.35	9.58	9.05	8.66	8.35	8.12	7.92	7.63	7.32	7.01	6.85	6.68	6.52	6.35	6.17	6.00
12	18.64	12.97	10.80	9.63	8.89	8.38	8.00	7.71	7.48	7.29	7.00	6.71	6.40	6.25	6.09	5.93	5.76	5.59	5.42
13	17.81	12.31	10.21	9.07	8.35	7.86	7.49	7.21	6.98	6.80	6.52	6.23	5.93	5.78	5.63	5.47	5.30	5.14	4.97
14	17.14	11.78	9.73	8.62	7.92	7.43	7.08	6.80	6.58	6.40	6.13	5.85	5.56	5.41	5.25	5.10	4.94	4.77	4.60
15	16.59	11.34	9.34	8.25	7.57	7.09	6.74	6.47	6.26	6.08	5.81	5.54	5.25	5.10	4.95	4.80	4.64	4.47	4.31
16	16.12	10.97	9.00	7.94	7.27	6.81	6.46	6.19	5.98	5.81	5.55	5.27	4.99	4.85	4.70	4.54	4.39	4.23	4.06
17	15.72	10.66	8.73	7.68	7.02	6.56	6.22	5.96	5.75	5.58	5.32	5.05	4.78	4.63	4.48	4.33	4.18	4.02	3.85
18	15.38	10.39	8.49	7.46	6.81	6.35	6.02	5.76	5.56	5.39	5.13	4.87	4.59	4.45	4.30	4.15	4.00	3.84	3.67
19	15.08	10.16	8.28	7.26	6.62	6.18	5.85	5.59	5.39	5.22	4.97	4.70	4.43	4.29	4.14	3.99	3.84	3.68	3.51
20	14.82	9.95	8.10	7.10	6.46	6.02	5.69	5.44	5.24	5.08	4.82	4.56	4.29	4.15	4.00	3.86	3.70	3.54	3.38
21	14.59	9.77	7.94	6.95	6.32	5.88	5.56	5.31	5.11	4.95	4.70	4.44	4.17	4.03	3.88	3.74	3.58	3.42	3.26
22	14.38	9.61	7.80	6.81	6.19	5.76	5.44	5.19	4.99	4.83	4.58	4.33	4.06	3.92	3.78	3.63	3.48	3.32	3.15
23	14.19	9.47	7.67	6.69	6.08	5.65	5.33	5.09	4.89	4.73	4.48	4.23	3.96	3.82	3.68	3.53	3.38	3.22	3.05
24	14.03	9.34	7.55	6.59	5.98	5.55	5.23	4.99	4.80	4.64	4.39	4.14	3.87	3.74	3.59	3.45	3.29	3.14	2.97
25	13.88	9.22	7.45	6.49	5.88	5.46	5.15	4.91	4.71	4.56	4.31	4.06	3.79	3.66	3.52	3.37	3.22	3.06	2.89
26	13.74	9.12	7.36	6.41	5.80	5.38	5.07	4.83	4.64	4.48	4.24	3.99	3.72	3.59	3.44	3.30	3.15	2.99	2.82
27	13.61	9.02	7.27	6.33	5.73	5.31	5.00	4.76	4.57	4.41	4.17	3.92	3.66	3.52	3.38	3.23	3.08	2.92	2.75
28	13.50	8.93	7.19	6.25	5.66	5.25	4.93	4.69	4.50	4.35	4.11	3.86	3.60	3.46	3.32	3.18	3.02	2.86	2.69
29	13.39	8.85	7.12	6.19	5.59	5.18	4.87	4.64	4.45	4.29	4.05	3.80	3.54	3.41	3.27	3.12	2.97	2.81	2.64
30	13.29	8.77	7.05	6.12	5.53	5.12	4.82	4.58	4.39	4.24	4.00	3.75	3.49	3.36	3.22	3.07	2.92	2.76	2.59
40	12.61	8.25	6.60	5.70	5.13	4.73	4.44	4.21	4.02	3.87	3.64	3.40	3.15	3.01	2.87	2.73	2.57	2.41	2.23
60	11.97	7.76	6.17	5.31	4.76	4.37	4.09	3.87	3.69	3.54	3.31	3.08	2.83	2.69	2.55	2.41	2.25	2.08	1.89
120	11.38	7.32	5.79	4.95	4.42	4.04	3.77	3.55	3.38	3.24	3.02	2.78	2.53	2.40	2.26	2.11	1.95	1.76	1.54
∞	10.83	6.91	5.42	4.62	4.10	3.74	3.47	3.27	3.10	2.96	2.74	2.51	2.27	2.13	1.99	1.84	1.66	1.45	1.00

[a] Multiply these entries by 100.

Table A.8 The Chi-Square Distribution

Values of $\chi^2_{\nu,P}$ such that

$$P = \frac{1}{2^{\nu/2}\Gamma(\nu/2)} \int_0^{\chi^2_{\nu,P}} y^{\nu/2-1} e^{-y/2}\, dy$$

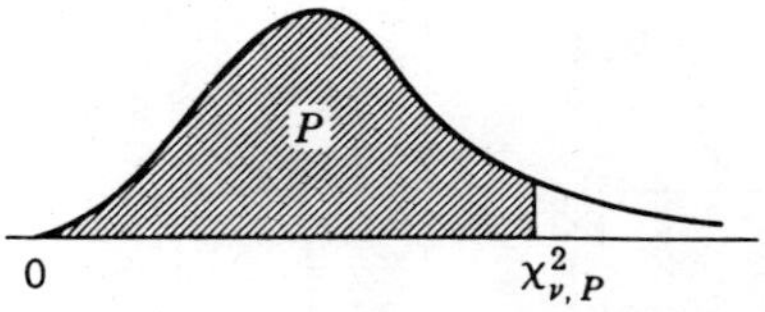

ν \ P	0.005	0.010	0.025	0.050	0.100	0.250	0.500
1	0.00004	0.00016	0.00098	0.00393	0.01579	0.1015	0.4549
2	0.0100	0.0201	0.0506	0.1026	0.2107	0.5754	1.386
3	0.0717	0.1148	0.2158	0.3518	0.5844	1.213	2.366
4	0.2070	0.2971	0.4844	0.7107	1.064	1.923	3.357
5	0.4117	0.5543	0.8312	1.145	1.610	2.675	4.351
6	0.6757	0.8721	1.2373	1.635	2.204	3.455	5.348
7	0.9893	1.239	1.690	2.167	2.833	4.255	6.346
8	1.344	1.646	2.180	2.733	3.490	5.071	7.344
9	1.735	2.088	2.700	3.325	4.168	5.899	8.343
10	2.156	2.558	3.247	3.940	4.865	6.737	9.342
11	2.603	3.053	3.816	4.575	5.578	7.584	10.34
12	3.074	3.571	4.404	5.226	6.304	8.438	11.34
13	3.565	4.107	5.009	5.892	7.041	9.299	12.34
14	4.075	4.660	5.629	6.571	7.790	10.17	13.34
15	4.601	5.229	6.262	7.261	8.547	11.04	14.34
16	5.142	5.812	6.908	7.962	9.312	11.91	15.34
17	5.697	6.408	7.564	8.672	10.09	12.79	16.34
18	6.265	7.015	8.231	9.390	10.86	13.68	17.34
19	6.844	7.633	8.907	10.12	11.65	14.56	18.34
20	7.434	8.260	9.591	10.85	12.44	15.45	19.34
21	8.034	8.897	10.28	11.59	13.24	16.34	20.34
22	8.643	9.542	10.98	12.34	14.04	17.24	21.34
23	9.260	10.20	11.69	13.09	14.85	18.14	22.34
24	9.886	10.86	12.40	13.85	15.66	19.04	23.34
25	10.52	11.52	13.12	14.61	16.47	19.94	24.34
26	11.16	12.20	13.84	15.38	17.29	20.84	25.34
27	11.81	12.88	14.57	16.15	18.11	21.75	26.34
28	12.46	13.56	15.31	16.93	18.94	22.66	27.34
29	13.12	14.26	16.05	17.71	19.77	23.57	28.34
30	13.79	14.95	16.79	18.49	20.60	24.48	29.34
40	20.71	22.16	24.43	26.51	29.05	33.66	39.34
50	27.99	29.71	32.36	34.76	37.69	42.94	49.33
60	35.53	37.48	40.48	43.19	46.46	52.29	59.33
70	43.28	45.44	48.76	51.74	55.33	61.70	69.33
80	51.17	53.54	57.15	60.39	64.28	71.14	79.33
90	59.20	61.75	65.65	69.13	73.29	80.62	89.33
100	67.33	70.06	74.22	77.93	82.36	90.13	99.33

Source. Reprinted from Table E, *Statistics and Experimental Design in Engineering and the Physical Sciences* by N. L. Johnson and F. C. Leone. Copyright © 1964 by John Wiley & Sons, Inc., New York. Also reprinted from *Biometrika Tables for Statisticians*, Vol. 1, by E. S. Pearson and H. O. Hartley, pp. 130–131. Copyright © 1958. Reprinted by permission of John Wiley & Sons, Inc. and the Biometrika Trustees.

Table A.8 *(Continued)*

v \ P	0.750	0.900	0.950	0.975	0.990	0.995	0.999
1	1.323	2.706	3.841	5.024	6.635	7.879	10.83
2	2.773	4.605	5.991	7.378	9.210	10.60	13.82
3	4.108	6.251	7.815	9.348	11.34	12.84	16.27
4	5.385	7.779	9.488	11.14	13.28	14.86	18.47
5	6.626	9.236	11.07	12.83	15.09	16.75	20.52
6	7.841	10.64	12.59	14.45	16.81	18.55	22.46
7	9.037	12.02	14.07	16.01	18.48	20.28	24.32
8	10.22	13.36	15.51	17.53	20.09	21.96	26.12
9	11.39	14.68	16.92	19.02	21.67	23.59	27.88
10	12.55	15.99	18.31	20.48	23.21	25.19	29.59
11	13.70	17.28	19.68	21.92	24.72	26.76	31.26
12	14.85	18.55	21.03	23.34	26.22	28.30	32.91
13	15.98	19.81	22.36	24.74	27.69	29.82	34.53
14	17.12	21.06	23.68	26.12	29.14	31.32	36.12
15	18.25	22.31	25.00	27.49	30.58	32.80	37.70
16	19.37	23.54	26.30	28.85	32.00	34.27	39.25
17	20.49	24.77	27.59	30.19	33.41	35.72	40.79
18	21.60	25.99	28.87	31.53	34.81	37.16	42.31
19	22.72	27.20	30.14	32.85	36.19	38.58	43.82
20	23.83	28.41	31.41	34.17	37.57	40.00	45.32
21	24.93	29.62	32.67	35.48	38.93	41.40	46.80
22	26.04	30.81	33.92	36.78	40.29	42.80	48.27
23	27.14	32.01	35.17	38.08	41.64	44.18	49.73
24	28.24	33.20	36.42	39.36	42.98	45.56	51.18
25	29.34	34.38	37.65	40.65	44.31	46.93	52.62
26	30.43	35.56	38.89	41.92	45.64	48.29	54.05
27	31.53	36.74	40.11	43.19	46.96	49.64	55.48
28	32.62	37.92	41.34	44.46	48.28	50.99	56.89
29	33.71	39.09	42.56	45.72	49.59	52.34	58.30
30	34.80	40.26	43.77	46.98	50.89	53.67	59.70
40	45.62	51.80	55.76	59.34	63.69	66.77	73.40
50	56.33	63.17	67.50	71.42	76.15	79.49	86.66
60	66.98	74.40	79.08	83.30	88.38	91.95	99.61
70	77.58	85.53	90.53	95.02	100.4	104.2	112.3
80	88.13	96.58	101.9	106.6	112.3	116.3	124.8
90	98.65	107.6	113.1	118.1	124.1	128.3	137.2
100	109.1	118.5	124.3	129.6	135.8	140.2	149.4

INDEX